Nov. 30, 02

CRYSTALS AND LIFE

KEITON,

My best wishes

for your future

at Abbott. May

your years be

enjoyable & productive,

but also happy

CRYSTALS AND LIFE

A Personal Journey

Cele Abad-Zapatero

INTERNATIONAL UNIVERSITY LINE
La Jolla, California

Library of Congress Cataloging-in-Publication Data

Abad Zapatero, Celerino.
    Crystals and Life: A Personal Journey / Cele Abad-Zapatero
        p. ; cm.
        Includes bibliographical references and index.
        ISBN 0-9720774-0-5 (pbk)
        1. Crystallography. 2. Crystals. I. Title.

    QD905.2 .A23 2002
    548--dc21

                                              2002011972
                                                    CIP

© International University Line, 2002
Post Office Box 2525,
La Jolla, CA 92038-2525, USA

Library of Congress Catalog Card Number 2002011972

Printed in the United States of America

10 9 8 7 6 5 4 3 2 1

ISBN   0-9720774-0-5 $24.95   Softcover

Dedication

No labor-saving machine,
Nor discovery have I made,
Nor will I be able to leave behind me any
  wealthy bequest to found a hospital or
  library
Nor reminiscence of any deed of courage
  for America,
Nor literary success nor intellect, nor book
  for the book-shelf,
But a few carols vibrating through the air
  I leave
For comrades and lovers.

Walt Whitman
(poem: *No labor-saving machine*, from
Leaves of Grass)

Contents

Foreword

Crystals have been a fascination to all peoples at all times. Crystalline precious stones, such as diamonds, sapphires and rubies, have been the basis of tradition, myths, high finance, exploration and wars since the beginning of recorded history. Nevertheless the systematic study of crystals, a subject known as "crystallography", is often considered as a boring and dry occupation, pursued by people who have largely lost touch with the affairs of the World. *Crystals and Life* bridges this gap in a beautiful way. The author, Celerino ('Cele') Abad-Zapatero, weaves his own life experiences together with emotions such as crying and dying, biographical sketches of philosophers and scientists, as well as observations about common objects such as sheets of stamps. He includes poems and ballads written by him in which he recalls moments of discovery and other events that have formed him and impacted many others. But this book is by no means an autobiography. It is a textbook that follows almost the same outline as the recently published *International Tables for Crystallography*, Volume F, produced by the International Union of Crystallography and edited by Edward Arnold and myself. The difference is that *Crystals and Life* radiates pleasure and fascination of the Universe in which we live, whereas Volume F is mostly stripped of this humanity.

Cele was born and grew up in Northwest Spain, near the city of Burgos, within the province of the same name. Even today I remember well this town and its large cathedral, as it was here that I first started to fall in love with Audrey, one of the girls in our group of young people traveling together around Spain in the summer of 1953. We both still remember the beauty of the country, but also the extreme poverty of its people. It is clear that Cele's inspirations for the present book came in part from his memories of Spain mixed with his experiences as a graduate student in Texas, as a post-doctoral fellow in my laboratory, working as a professional crystallographer at Abbott Laboratories, and raising a family in a suburb of Chicago. The variety of Cele's background and the richness of his English make this book into a unique blend of literature and scientific education.

It has been my privilege to host and work with many pre- and post-doctoral students with vastly different cultural backgrounds during almost forty years at Purdue University in Indiana. We have together enjoyed the pleasures of discovery and agonized over disappointments. Cele's book is about these emotions and the people who have experienced them.

Michael G. Rossmann
Department of Biological Science
Purdue University
West Lafayette, Indiana

$\mathcal{P}$reface

The majority of these brief essays were published in professional newsletters of the American Crystallographic Association (ACA), the International Union of Crystallography (IUCr) or the Protein Data Bank (PDB) Quarterly Newsletter under the heading *Notes of a Protein Crystallographer*. Eventually, it occurred to me that I could make a self-contained unit if I added a few more essays covering the foundations of crystallography in a non-technical manner. These pages, properly revised and illustrated, could communicate the discoveries and wonders of crystallography to a wider and younger audience, as well as that convenient abstraction that we refer to as the "educated layperson." Crystallography, as a science or as a field of research, has a dreadful reputation. It is always shrouded in a veil of mathematical mystery that apparently only a few can lift to discover its innate beauty. Although the professional crystallographer needs a mathematical background to practice the craft, this does not mean that complex mathematical knowledge is required to understand what crystallographers do. Appreciation and further interest may follow an initial window of understanding into the field.

There was one further motivation. I have read many books of science addressed to the layperson and I have found many of them terribly dry. A sentence-after-sentence recitation or regurgitation of facts, only broken with a few parentheses to mention who obtained the Nobel Prize for what and on which date. Nowadays, the availability of Internet sources has made such an approach to convey the scientific enterprise obsolete. We need to present how rich and multifaceted the scientific endeavor is in a different manner: peripatetic, colloquial, narrative, dramatic, literary and even poetic. Roald Hoffman, Carl Djerassi, Oliver Sacks and others have made major strides along these lines within the domain of chemistry. However, crystallography, hidden behind its mathematical cloak, is still among the orphan sciences in this regard. So far, it seems that it is better to leave it alone.

The fundamental discoveries and the names with the appropriate accolades can be obtained from a myriad of sites in the world wide web. Can we, however, convey the basis of crystallography in a condensed, personal, suggestive, inspirational and poetic form? Although it is a challenge, I think that the answer should be "Yes." Rachel Carson, the influential author of *Silent Spring*, wrote in a 1956 article:

> "Once the emotions have been aroused-a sense of the beautiful, the excitement of the new and the unknown, a feeling of sympathy, pity, admiration or love—then we wish for knowledge about the object of our emotional response. Once found, it has lasting meaning."[1]

She wrote these words in an article about exploring nature with children but all through her books and articles we can sense these emotional reverberations with the world she was describing and explaining to the reader. In our technological world, where all the facts are literally at our fingertips, we have to awaken anew a sense of curiosity, wonder and the interconnection between the sciences and the arts, among children, adolescents and adults alike.

Those are the goals of this collection of vignettes around the general theme of crystals, crystallography and the implications of their scientific findings for the understanding of life processes. Because of my professional expertise, I have focused on the study of crystals from biological samples, and especially proteins. However, enough material is common to many other applications of crystallography that the reader should be able to extend the ideas easily. A limited glossary has been added at the end to provide the definitions of some of the technical terms used throughout the text and there are notes and references for further reading. A brief list of World Wide Web sites of interests has also been added at the end.

On one side, the crystallographic community has failed to communicate the enormous influence that the study of crystals, and the study of matter inside crystals or semi-crystalline materials, has had on our tangible world. On the other hand, the public at large has not realized how the discoveries made in the atomic domain of crystals are affecting their daily lives in areas such as health care or the design of new materials. Moreover, the biomedical sciences stand at the brink of yet another revolution fueled by the influx of information obtained from the unraveling of the gene maps of pathogens and the human genome itself. Mapping the three-dimensional structures of a large portion of the proteins coded by those genes is the next challenge in molecular biology. This knowledge will be followed by therapies based on the understanding, inhibition or alteration of the three-dimensional structures of the proteins associated with those genes. How are those structures unveiled? What is the technology behind those spectacular computer images? What do they mean for the synthesis of novel drugs, or the design of new vaccines? What can we expect in the near future? What are the implications of this wealth of new information for our understanding of life's physico-chemical processes?

Reading through these pages, the reader should be able to assimilate a series of ideas and concepts related to crystallography. She or he will be able to conceptually understand how crystallography works and how crystallographers are able to unveil the molecular structures existing within crystals. Although painted with coarse strokes, there is enough historical and conceptual background dispersed throughout the text to follow the development of the field and to provide a firm basis for further study. Some essays are also meant to identify and pay homage to some of the heroes of the field who developed the concepts, methods and tools to expedite the structural analysis. A few articles will also illustrate how results obtained from crystals or crystalline materials are affecting our standard of living and quality of life in certain specific areas. Finally a group of essays is devoted to issues that are related to the future of structural biology such as what are the possibilities of this technology, which areas of the field and thriving, and what can we expect in the years ahead.

After a brief personal evocation (Part I), the individual essays are broadly grouped under seven additional Parts (II–VIII): "The Basic Elements of Crystallography" (Chapters 1–4); "Symmetry and Properties of Protein Crystals" (5–7); "From Data to Electron Density Maps" (8–11); "Protein Structure, Model Building, and Refinement" (12–14); "New Technologies" (15–18); "Practical Applications" (19–22); and "Future Perspectives" (23–27).

These personal and scientific sketches can be used as an introductory narrative to complement courses in protein crystallography or structural biology. They can also be used to introduce the fields to non-science majors. However, any person curious about crystallography and its impact on the biomedical sciences and our world at large can also benefit from reading them. They can be read in sequence from beginning to end or at random. Like poems, some essays are easier to read than others, but I encourage you to make an effort to read through all of them. If you already know crystallography, I suggest that you pick and choose first. If you are a novice to the field, you should begin reading Parts II–IV and come back to them as needed. Some of the essays are intended to be read like a poem, in a coffee break; others on a quiet Sunday brunch at home or on the train or bus. All can be read anywhere and everywhere. If I succeed in brightening your day with an "*Aha!*" of scientific or human understanding, these vignettes will have served their purpose.

$\mathcal{A}$cknowledgements

his book is truly a collective endeavor. The scientific discoveries and the factual information presented, described, explained and narrated within these pages resulted from the labor of thousands of individuals. This is the unique and collective character of the scientific adventure; crystallography is no exception. My contribution might be that of telling the story in my own way, woven with my own personal experiences. In writing these lines, quite often I have felt like a balladeer that was narrating the foundations of crystallography, its achievements, applications and its frontiers to a larger interested audience of non-crystallographers. Thus, conceptually at least, I should acknowledge the worldwide community of crystallographers for allowing me to write about their discoveries reported at scientific meetings, published or simply shared with me over a cup of coffee or during a congenial lunch. I would like to do so here, wholeheartedly. Special thanks also to Profs. Larry Fox, Marv L. Hackert and Hugo Steinfink, my Ph.D. mentors at UT Austin; without them, this adventure would have probably never taken place. None of this could have been written without my wonderful years under the guidance of Prof. Michael G. Rossmann at Purdue University.

In addition, many friends and colleagues have seen this narrative grow from a collage of disjointed essays to a more unified opus thanks to their readings of previous drafts and to their ensuing comments, suggestions and criticisms. These closer participants in the project are many and I would like to mention who they are. I apologize in advance for any omissions. Terese M. Bergfors, Cathy Brennan, Mark Bures, Bill Duax, Ignasi Fita, Federico Gago, David S. Goodsell, Eric Hebert, Deanna Haasch, Ignacio Hernando, Osnat Herzberg, Pat Lunn, Javier Mazana, Ana Pereda López, Andrew Petros, Luis M. Quirós, Rosario Recacha, Ian K. Robinson, Tim Rydel, José Luis San Emeterio, Ben Stein, and Kent Stewart read my earlier versions.

I also thank Joel Sussman and the staff of the Brookhaven PDB for printing earlier versions of many of these essays. Ron Stenkamp, Judith Flippen-Anderson and Marcia Evans, editors of the *ACA Newsletter*, published some of these essays. Susan Strasser from the APS office provided materials. I received comments, support, and encouragement from Héctor and Cecilia Rasgado Flores, and colleagues from the *Sociedad de Biofísicos Latino Americanos* (SOBLA), Mark Hermodson, William McClelan, Ivan Rayment, Theresa Sylvester (from the CLS in Canada), Karl Walter and Herman Winick.

Undoubtedly, certain people have played a more significant role in the development of this manuscript. David S. Goodsell gave me sound advice. Ms. Joyce Gatto polished my knowledge of English as a Second Language at the College of Lake County (Grayslake, IL). Profs. Phil Coppens and Jenny Glusker directed me towards the Book Committee of the IUCr, which encouraged me in the early stages of the project. Prof. Glusker read several drafts of the manuscript and always offered invaluable editorial advice. My Internet friend Vivienne B. Gerritsen from the *Institut Suisse de Bioinformatique* (ISB or SIB for the English speaking world) in Geneva, Switzerland, whom I never met in person, read some of the original essays and an early version of the manuscript, and always provided me with insightful suggestions and superb language advice. Gale Rhodes has guided me through the pains of authorship and publication. Sharon Wilder, from Purdue University was a bottomless resource of information. Ms. Kyle D. Burkhardt gave friendship, constant support and inspiration. Since crystallography is a visual science *par excellence*, this book will not be what it is now without the unfailing and superb technical assistance of the people at Abbott Laboratories' Creative Network Department through many years, especially Jeff Frye and Jim Sukowski.

Thanks to Mr. and Mrs. Greub of Gerhard's in Lake Forest for their weekly masterpiece, French *brioche*, and for serving me at odd hours on weekends during the development of this work. Lake Forest Library was a hospitable institution; its kind and friendly staff members allowed me to browse the new book section an innumerable number of hours. The reference section was patient when I asked them to locate the most obscure quotations or to loan rare books, particularly Ms. Mirdza E. Berzins. I owe thanks also to the Lake Forest Book Store, for their patience and willingness to help in innumerable occasions. I must also mention my department manager for many years at Abbott Laboratories (Dr. J. Greer), our Advanced Technology area head (Dr. Jim Summers), and the vice-president of research in the pharmaceutical products division, Dr. Dan Norbeck, for allowing me to use the resources and facilities of Abbott Laboratories during the creation of this personal work.

I would like to express my deep gratitude to Igor Tsilgeny, Laura DeCooze, Nina Bogucharoff, and the staff at IUL for their diligence and competence in bringing this dream into reality, as well as to the scientific editor Lynn Ten Eyck for gentle but invaluable suggestions.

Finally, the role played by my most immediate family, my wife and our two children, is woven throughout the text as is their intangible contribution and my infinite debt to them. Similarly the implicit gratitude to my late parents Juan Abad and Amparo Zapatero permeates the book. What I owe to all of them is more than I can express in a few short sentences here.

*L*ist of *I*llustrations

Figures

Color Plates

1

brief evocation: crystal reminiscences

The time is December 1999, Christmas Eve before the new millennium. I am rapidly wrapping presents for our two children. For the past few years, I have been trying to include a spe-cial natural object as a present along with the standard Christmas list of today's teenagers. Last year it was a cluster of milkweed seeds, still packed in their leathery case. The year before, it was a beautiful pinecone. Prior to that it was some unusual maple seeds. I am just trying to impress upon them that besides the fashionable and frequently expensive objects from our society one can find beautiful, unique and free objects in our parks and sidewalks. But this year it has to be something very, very special. Finally, the inspiration comes. I rush to the garage, and from the bottom shelf I pull a heavy shoebox with the interior still wrapped in an old Spanish newspaper from the early 1970s. This is a very special box indeed that crossed the Atlantic containing many precious items.

As I unwrap the newspaper, my spirit travels across the ocean to the hills of the Castillian plateau, near Valladolid in northern Spain. My heart rejuvenates by approximately thirty years and I see myself climbing up a steep hill, hand-in-hand with my wife-to-be Vithorita (in Basque, familiar diminutive for Victoria), sweating under the simmering sun of the treeless landscape but excited and happy. During our college years and on certain weekends, she and I would make long hiking excursions with our modest means to the dusty hills surrounding Valladolid searching for minerals and fossils. Our equipment was only a small backpack with some sandwiches, and a poor geologist's hammer that we purchased very inexpensively. The landscape is almost barren, with sparse vegetation at both sides of the riverbed. The hills are chalky white.

As we climb up, we can recognize the brutal effects of the free running water of the storms on a landscape deprived of vegetation. Yet, gathered in the crevices and irregularities of the terrain we can see a myriad of large, rough, dusty, scratched gypsum crystals. Those crystals combine in pairs to give the appearance of arrowheads; exactly as the textbooks say about gypsum crystals. We can not believe our eyes. We never expected such a treasure of crystals in such a depressing landscape. We collect the best specimens of several types and full of excitement put them in our backpack.

As I contemplate and touch some of those natural artifacts, another excursion in the company of Vithorita and friends floats in mind. This one is closer to my hometown—in the proximity of a town called "Campaspero" (contraction of *Campo áspero*: rough field) near the province of Burgos. Amazingly enough, this old village has sprung out on what appears to be a limestone quarry first exposed and then eroded by old torrents and rivers. At the entrance of the town, the terrain is truly rugged to the point of making it almost impossible to lay your shoe flat by the side of the road. We pass the center of the village and began searching on some of the cultivated fields near the river. Locals as well as friends have told us that every spring, when the fields are plowed, new fossils come up to the surface and can be gathered without much effort. We are not disappointed. We soon find all kinds of fossilized sea urchins, various kinds of shells from extinct snails, and clams of various sizes. The sea urchins are particularly well preserved as they still have the bumps on the surface that served as attachment points for the spicules.

Certainly, I have found the perfect present for our children for this very special Christmas: crystals and fossils that were found and collected by their parents before they were married. I quickly take one of each, wrap them in the original newspaper, and put them in two marmalade jars under the tree. Although not spectacular gemstones, gypsum crystals have soft, sparkling, surfaces that can be cleaved as layers with the fingernails. Like small, squashed, pumpkins these fossilized sea urchins preserve the exquisite detail of the external skeleton of long-extinct living organisms. I would like to convey and explain to our children, and you, the reader, the natural and hidden beauty that exists within crystals and even in the simplest and more unpretentious natural objects. I would like to show how crystals aid in unveiling beautiful forms and patterns in Nature. I will try to make it congenial, like a friendly dialog. Do you wish to participate?

II

the basic elements
of crystallography

1

The Magic of Crystals: What Is a Crystal?

In the vernacular usage, the word "crystal" resonates with purity, perfection, transparency and even durability. In "*crystalore*", crystals are supposed to have special properties to energize and heal people. These exceptional properties of crystals are stretched to the limit in the expressions "crystal gazing" and "looking at the crystal ball" implying that crystals allow us to see into the future. From the human standpoint, a very special "crystal" and the most precious of the gemstones, diamond, is also associated with pure and faithful love. It might be that given the fragility and weaknesses of the human heart, we give diamonds—the hardest, most durable and most pristine of the gems—to ensure eternal and unfailing love or possibly to protect our most dear ones.

The word itself comes from the Greek (*krystallos*) and one can find several translations around the idea of "frozen ice." The Greeks invented the term because the early quartz specimens were believed to be water frozen by the intense cold of distant, almost inaccessible mountains. Most likely the remote sources of these specimens, produced by insurmountable forces in distant places, added to the mystery of these geometrical objects. Pliny the Elder was the first compiler of the scientific terminology of the Greeks, and perished in the eruption of Vesuvius (74 A.D.) recording those events. He wrote in his *Natural History*: "crystal is only found in those high places where the winter snows have gathered in great quantity, and it is surely ice;

and for this reason the Greek have given it its name." The geometric nature of these "crystals" was also described by Pliny when he talks about "*sexangulum*" in referring to the hexagonal cross-section of quartz, the mineral most commonly associated with the term "crystal."[1,2]

For simplicity, the conventional history that we learn during our educational years is typically narrated in terms of heroes and villains. Similarly, we can dichotomize the individuals that contributed to the development of science into visionaries and uninspired scientist; heroes and villains. Those qualifiers depend upon whether they proposed "clear vistas" or "dead alleys" respectively. The path that the early crystallographers traveled to unravel what crystals really are was not very different from the one followed by other branches of science. As more and more samples were found and characterized by their transparency, color, and especially geometrical regularity, the study of crystals faced the problem of getting a handle on the immense variety of crystal forms. A French ex-officer of the war of India (Jean-Baptiste Louis Romé de Lisle, 1736–1790) was to cut a clear trail through the jungle and introduce the right criteria to classify crystals.[3]

Romé de Lisle was imprisoned by the English for five years and eventually brought to France in 1764 as a free man. Upon his return he developed a passionate interest in the study of minerals, cataloguing many of the private collections of the time. The success of the classification of biological forms by Carl Von Maria Linnaeus (1707–1778) at about the same time lured him initially towards a classification of the crystalline forms based on their external appearance. This system however, was not successful in classifying the myriad of forms presented by crystals and specially in recognizing different forms of the same natural substance.*

In the history of science and technology, many of the unsung heroes are instrument makers (see, for instance, the delightful book *Longitude*

---

* A brief but beautifully illustrated booklet documenting 200 years of crystallography in France was published on the occasion of the XVth Congress of the International Union of Crystallography that took place in Bordeaux, France in July 1990. The information is available on the web linked to the web address of the Laboratory of Mineralogy and Crystallography of Paris: http://www.lmcp.jussieu.fr

by Dava Sovel).[4] We glorify the theoreticians who impose order on the kaleidoscope of shapes and forms with their platonic notions. However, we tend to ignore the craftsmen who created the instrument who made certain measurements possible; in due course, these measurements were critical to provide basis for the unifying idea. In this case, the manufacture and usage of a precision contact "goniometer" (i.e., instrument to measure angles accurately) by his assistant Arnould Carangeot (ca. 1780) allowed Romé de Lisle the careful measurement of the angles between the faces (dihedral angles) of various geometrical forms of the same crystal specimen. The description and manufacture of a more accurate reflective instrument by William H. Wollaston (published in 1809) permitted the measurements of dihedral angles in smaller crystals. This parameter did indeed permit Romé de Lisle to establish and enunciate the law of constancy of angles: "in all crystals of the *same* substance the angles between corresponding faces have the *same* value (italics are mine)." The measurement and comparison of these angular parameters among different crystals permitted the correct classification of the various crystalline forms and established, for the first time, a connection between the geometrical features of the crystals and the nature of the substance forming them. This insight allowed Romé de Lisle to make a clear distinction between two apparently similar, deep red, minerals, ruby and spinel, as two different precious stones. Thus, although following Linnaeus' he initially attempted to classify crystals based only on external shapes and forms, Romé de Lisle eventually found that it was the constancy of angles that proved to be essential to organize the universe of crystal forms.

The next hero in the history of crystallography is René-Just Haüy (1743–1822), the son of a poor French weaver who, like many in those times, was able to obtain an education only by joining the church. He studied the classics and the natural sciences of the time, namely botany and physics, reaching the level of abbot. Because of his religious affiliation he suffered persecution and was almost put to death during the French revolution. However, he was appointed a member of the Academy of Sciences during the reign of Napoleon and became recognized by scientific societies in France and all over

Europe because of his contributions to mineralogy and the study of crystals. Yet, his modest persona never abandoned him.[2,3]

Abbot Haüy discovered the study of crystals at the rather old age of thirty-five, most likely by accident. He unintentionally dropped (and naturally broke) a specimen of calcite crystals and noticed that the resulting fragments had the same shape and the same oblique angle as the original specimen. This is a commonplace observation for any person familiar with calcite crystals but in the case of Abbot Haüy, resulted in an idea that inspired all of his seminal contributions to crystallography. From this fortuitous observation, he concluded that crystals were built of a large number of smaller, simpler basic units, so small that the resulting faces of the crystal looked smooth, all of which had the same shape. This elementary unit he named "integrant element" or "integrant molecule." He refined this original hypothesis in two major works (*Traité de Mineralogie*, 1801; and *Treatise of Crystallography*, 1822), and proposed that a limited number of three basic building blocks are needed to construct a crystal, much like bricks making up a house.

The concept of identical repeating units forming a crystal provided the two critical elements, which define a crystalline lattice in modern terms: i) the underlying repetition or symmetry of a motif in a geometrical array of points (i.e., a lattice); ii) the requirement for the building block or "brick" to fill the space without internal voids or spaces. **Plate 1.1** illustrates the basic notion of a crystal using a familiar object, a sheet of stamps. The motif used to create a single stamp is repeated on the horizontal and vertical directions to create a two-dimensional crystal as formed by this ordinary object.

Although much more elaborate, similar patterns can be found on the design of wallpaper and fabrics. **Plate 1.2** provides a more artistic rendering of the same idea: a three-dimensional crystal is formed by the repetition of a motif (i.e., a fish) in three dimensions. In the perspective offered by the artist, it seems as though the viewer is inside the infinite array of fish that constitutes a crystal.

The history of crystallography continues with Auguste Bravais (1811–1863), a marine officer of insatiable curiosity who contributed to many areas of science (astronomy, botany, meteorology, scientific exploration and others).[3] Bravais established rigorously the geometric aspects of crystallography with the introduction of the fourteen unique minimal boxes or lattices, which could generate the forms observed in the seven crystal systems that are described below. The work of Bravais required the systematic examination of the various ways in which three-dimensional space can be filled without leaving any empty spaces. To visualize what he did, I can refer you to the different ways in which a floor can be tiled using only *one* type of tile throughout. It is common knowledge that only certain shapes can be used. Most commonly only square, rectangular and hexagonal mosaics are used. Bravais found that the basic building block of crystals could only be one of fourteen unique types (Bravais lattices). These lattices are normally grouped by increasing symmetry into seven crystal systems: triclinic (least symmetry), monoclinic, orthorhombic, tetragonal, trigonal, hexagonal and cubic, which has the highest symmetry (any textbook will illustrate these minimal unit cells, **Figure 1.1**). All this geometric information has been rigorously codified in a series of reference volumes called *The International Tables for Crystallography*.[5,6] This set of books is "the Bible" of crystallography and is consulted almost daily by crystallographers all over the world, either directly or encoded in the computer programs that they use.

Consistent with my Preface, I will not recite any more facts and names regarding the history of crystallography. Those can be found in the appropriate places. I will just return to the title of this essay and ask, "What is the magic of crystals?" What is their enchanting spell that fascinated humans from the moment the first crystals were discovered? Naturally, I do have an answer to this question. It is nothing so mundane as supernatural healing properties, which they do not have. Through these pages, I am going to let you browse, read, contemplate, ruminate, and explore the universe around crystals and molecular forms. I would like to be your guide during this expedition. You will come to your own conclusions. We can meet at the end of the journey and compare notes.

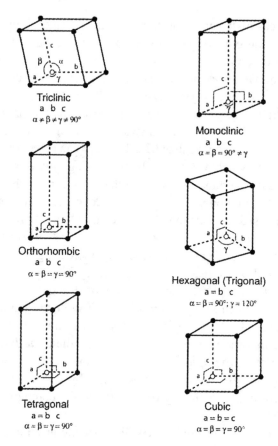

Figure 1.1. A schematic depiction of the simplest (or primitive) Bravais lattices selected from the unique fourteen geometric boxes (or cells) that can fill the three-dimensional space without gaps and generate all types of crystals. They are arranged in order of increasing regularity or symmetry from the triclinic (no symmetry: the three cell axes **a**, **b**, and **c** are different in size, and none of the angles among them is orthogonal: all different from 90°) to the cubic (highest symmetry: **a**=**b**=**c** and $\alpha=\beta=\gamma=90°$).

For consistency the monoclinic system is depicted with c as the unique axis. Although the rotational symmetry of the hexagonal (six-fold) and trigonal (three-fold) systems is different, the geometrical unit cell is the same and thus this cell is valid for both systems: i.e., a total of seven crystallographic systems.

2

These Naughty, Naughty X-rays: Properties of X-rays

The Roentgen Rays, The Roentgen Rays,
What is this craze;
The town's ablaze
With the new phrase
Of X-rays' ways
I'm full of daze
Shock and amaze,
For nowadays
I hear they'll gaze
Thro-cloak and gown-and even stays
These naughty, naughty Roentgen Rays.

(Published by the *Punch Magazine* in England at the time of the discovery of X-rays. Source: *Nature* (1995), special issue commemorating the one hundred anniversary of the discovery of X-rays)

X-rays are a widespread commodity today, from dentist's offices to hospitals, to airport security booths. However, when they were discovered in 1895 by Wilhelm Conrad Roentgen (or Röntgen), a German physicist (1845–1923), they caused quite a stir. For the first time many anatomical details of bones in the human body could be seen without any surgical invasion of the intact structure. Although considered to be completely inoffensive at the time, their damaging effects were soon also discovered as longer exposures became more common. Nowadays, we have become used to their penetrating power and their ability to go through "cloak and gown" revealing arteries, organs and tissues, as well as suspicious objects in our luggage. Of course, X-rays could and were used for many other practical applications, but it was their ability to make visible the structures behind the human skin that made them and Prof. Roentgen immediately famous (**Figure 2.1**).

From the viewpoint of their physical nature, the name given initially of "X-rays" (*X-Strahlen*)[1] was purposely ambiguous. They were assumed to be rays, like sunrays, but their true character was unknown. It was not possible to tell what this new physical entity was, whether it behaved like a wave perturbing the air and implying a connection with electromagnetic waves, or like a stream of minute balls striking a target, reminiscent of the corpuscular theory of light. The number of practical applications grew rapidly but their nature continued to puzzle the scientific community. They could penetrate the hardest of materials but they could not be bent like light by a transparent prism.

In the branch of science known as mineralogy, the human mind was constructing a scheme by which crystals were somehow built of clusters of atoms, still of unknown dimensions, repeated in three dimensions to create the

Figure 2.1. A caricature of Wilhem Conrad Roentgen, the German scientist that discovered X-rays, published in *Lustingen Blattern* circa 1900. Reprinted from *Nature*.[2] Courtesy of Macmillan Publishers Ltd.

amazing variety of crystalline forms. In the interaction of matter (crystals) with radiation (light or X-rays) the relative sizes of their constituents is very important because if there is great disparity between the two, the radiation passes through matter undisturbed. This is what was observed with most of the experiments with X-rays.

Max von Laue (German physicist, 1879–1960) is our next hero. Using the basic notions of chemistry at the time, he conjectured that the inter-atomic spaces in crystals were of the order of 10^{-8} cm (0.00000001 cm; 1 inch = 2.54 cm). Moreover, he argued that if the estimated wavelength (i.e., separation between two exactly equiva-

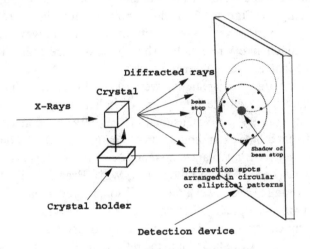

Figure 2.2. Schematic representation of an experiment involving the diffraction of X-rays by a crystal. Incoming radiation from the left is indicated by an arrow labeled *X-rays*. A single crystal is held on a crystal holder that allows the rotation of the crystal while being exposed to the X-rays. Multiple arrows emanating from the crystal indicate diffracted rays. The detector is represented by a rectangular surface on which the diffracting X-rays are recorded. These rays appear on the detector as spots typically arranged in circular or elliptical patterns. A small circle placed after the crystal in the diagram is a small "beam stop" or trap for the non-diffracted rays. This is typically a very small piece of lead that absorbs the X-rays emerging from the original beam. Due to its presence, a lighter shadow appears at the center of all the diffraction patterns (see **Figure 4.1**).

lent points in a wave) was of the order of 10^{-9} (smaller by a factor of 10 or so), it would be possible for a crystal to diffract X-rays. Diffraction means that the interference among the different outgoing waves results only in a few wave directions enhancing each other. The waves in other directions cancel out. Prof. von Laue discussed the possible experiment with two of his research students, W. Friedrich and P. Knipping, and after several unsuccessful attempts, they succeeded in obtaining what is known as a "diffraction pattern" from a crystal of copper sulfate. The experiment involved selecting a crystal, putting it in front of a source of X-rays, and placing a photographic plate behind the crystal. The result was a series of discrete and distinct spots arranged in circles or ellipses (**Figure 2.2**). I should point out that the source that they used was a mixture of X-rays with many different wavelengths or colors, much like white light. These results were published in 1912, more than ten years after the first discovery of X-rays.

The early diffraction patterns were very difficult to interpret because they involved the interaction of radiation of an unknown wavelength with a three-dimensional crystal. Crystals consisted of many different atoms extending in three-dimensions, which could be thought of as forming planes in many directions in space. Two major pioneer heroes of crystallography, W. H. Bragg (1862–1942) and W. L. Bragg (1890–1971), father and son, played a major role in simplifying this complexity. The younger had the insight of reducing this three-dimensional puzzle of the interaction of crystals with X-rays to two dimensions, by reducing it to the reflection of X-rays by a series of planes of equal inter-planar separation. He reasoned that X-rays reflected from a series of planes separated by a constant distance **d** will only add up with each other if the difference in their respective paths is equal to a multiple of the wavelength λ of the X-rays (analogous to the color of visible light). Thus

$$\text{difference in path} = \mathbf{n} \times \lambda,$$

where **n** is an integer number.

The resulting equation, known as **Bragg's Law** ($2\,\mathbf{d}\,\sin\theta = \mathbf{n}\,\lambda$), is one of the cornerstones of crystallography (see **Glossary**).

I refer you back to **Plate 1.1** where a crystal is simulated in two dimensions by a sheet of stamps. Some of the planes that the younger Bragg imagined can be easily visualized on our planar crystal by drawing lines that intersect the horizontal (**a**) and vertical (**b**) axes in a consistent manner. For example, I can draw an imaginary plane by connecting the horizontal (one unit East) and vertical (one unit North) end-points of the first stamp with a line; naturally, this will be labeled the (11) plane. Drawing similar lines through equivalent points two, three, four, etc. units away will give a set of parallel planes with a constant separation between them (planes (22), (33), (44), etc.). In a three-dimensional crystal, these planes will be described as (110) (**Plate 1.1**; green lines) since none of them intersect the missing third axis, which would come out of the figure. The labels of general planes having non-equal indices (for example, 130 or 120) can be obtained using simple rules described in the textbooks[3] and are illustrated in **Plate 1.1** and later in **Plate 3.1** with blue and red lines, respectively. Stop and think for a moment. You will soon realize that any series of planes having the same distance between them can be described in three-dimensional space by two parameters: their direction as given by a perpendicular (or *normal*) to the planes (as in the thin green lines with arrowheads in **Plate 1.1**), and the common distance between successive planes.

Crystal planes and faces had been described previously by mineralogists by assigning three integer numbers h,k,l to their direction in space. These three numbers were known as the Miller indices (W. H. Miller, Welsh crystallographer, 1801–1880). Now you can have a sense of the connection between the geometrical description of the crystals as an array of atoms, and the parameters used by the traditional mineralogists to describe planes and faces in real crystal specimens. Using this geometrical simplification the younger Bragg was able to explain the spatial arrangement of atoms in common salt (sodium chloride, NaCl), and many other simple chemicals. X-rays could now be used to unveil the hidden atomic structure of the crystals and to obtain quantitative information as to their internal distances and building parameters.

I know it has been very long, and I have almost violated my promise of not reciting many facts. I did not have any other choice. I'll try to amend the situation. Think about the ramifications of these experiments that were performed within a short time. In the field of physics, a new form of radiation of unknown nature was found. Mineralogists, faced with explaining the properties of crystals, had come to the conclusion that they had to be made of geometric building blocks that contained atoms arranged in still unknown ways. The experiment of Max von Laue and colleagues is the one-of-a-kind experiment where one kills five birds with one stone. After all the dust was settled, humankind was able to: i) establish the wave nature of X-rays because they can be diffracted by crystals; ii) analyze the pattern of diffracted rays, and draw in three dimensions how the atoms are arranged in the crystal; iii) begin to understand the rules of how the atoms combine in solid matter; iv) obtain exact dimensions for the distances of atoms and planes in the crystal and the value for the wavelength of the X-rays; and v) utilize this penetrating radiation to probe into the atomic structure of matter. Thanks to our son Pablo, I am beginning to understand a little bit of baseball. It seems to me that Laue's experiment is like a "grand slam" of scientific insight.

You have done the work. Now, let your imagination fly. What is the arrangement of atoms in a seashell or in the bark or wood of different trees? What molecular gymnastics make muscle contraction possible? What is the molecular structure of wonder drugs such as penicillin and how do they work? What makes silicon such a special material for electronic applications? At the atomic scale, how do biological processes work? How do molecules regulate our moods and daily cycles? Which atomic mazes must light particles travel in the leaves of plants in order to drive the critical chemical reactions in the biosphere? What atomic contacts make possible for viruses to infect the host cells and reproduce within them? What can we do to prevent it? Which molecular perturbations are responsible for certain types of cancer? Can we alter the atomic structure of silk to manufacture better natural or artificial fibers? What is the molecular design of a seagull feather? How can we imitate the lightness of a dandelion seed?

What molecular events imagine, design and paint the exquisite velvet of a flower petal? What atomic tapestry prevents us from bleeding to death when we cut ourselves? You can continue on with your own questions. Experiments using the diffraction of X-rays by crystals or quasi-crystalline samples are being performed all over the world to address those questions and many more at the atomic scale. Crystallographers, as the practitioners of the craft of crystallography, are in the middle of all these scientific and technical developments, using ever more intense X-rays and more sensitive detectors on forever more ingenious diffraction experiments.

3

Spatial Counterpoint: Real and Reciprocal Lattices

In musical parlance, counterpoint is the art of combining simultaneously two melodic lines in a composition. Although the term is somewhat related to polyphony, they are two distinct musical concepts. Polyphony or polyphonic, as exemplified in magnificent masses of the 16th century composer Giovanni Pierluigi da Palestrina (1525–1594), refers strictly to music that contains many different voices simultaneously. Counterpoint, on the other hand, is a term reserved for the craft or technique required for handling those two or more melodic lines. It is normally accepted that good counterpoint technique in a composer implies (1) some degree of independence between the melodies, or in other words a certain balance in the "horizontal direction"; and (2) harmonious interactions and relations between the different lines, or "vertical" concord.

In the history of musical composition, Palestrina is considered to be the premier exponent of polyphonic music, while J. S. Bach represents the zenith of counterpoint. After the codification of musical tonality presented by Bach in his two books of *The Well Tempered Clavier* (1722, 1742) many of Bach's compositions, including *The Art of Fugue* (1745–1750), are rich in contrapuntal wizardry. A musical theme is presented and developed, followed by a second or even third melodic line in a harmonious tapestry according to strict rules of tonality and counterpoint. In this context, I should point out that during the last decade of his life J. S. Bach was a member of a learned musical society called the Society of Musical Sciences, and it has been documented that several of his last works (i.e., *Canonic Variations*) were presented to the members of the society for distribution and study.[1]

In the domain of the graphical arts it is also possible to find examples of two different images sharing the attention of the viewer interwoven on the same canvas. This kind of experimentation was very popular among artists exploring the basis of visual perception in the early 1920s. Edgar Rubin's famous image of a white vase on a dark background (or could it be two dark faces on a white background?), is a well known example. I am no expert in the area and so I will only direct the reader to a new and unique organization dedicated to promote cross-disciplinary research in the arts and sciences. The Art Science Research Laboratory, Inc. (ASRL) founded in 1998 by Rhonda Roland Shearer and Stephen Jay Gould[*] is an amazing institution dedicated to the documentation, study and dissemination of themes in the *artsciences* realm.[2]

Although I will have more to say about the graphics work of M. C. Escher (1898–1972) later, I would like to select his work now to illustrate the presence of two simultaneous motifs on a surface, and to highlight a persistent theme underlying many of his lithographs and carvings. I am referring to the presence of images of good and evil as portrayals of order and chaos. Angels and devils share the regular division of the plane on a drawing that Escher executed with many different techniques,[3] and that he later used in the woodcut *Circle Limit IV*[4] and on a sphere carving in 1942. In other cases, he drew peaceful birds in the presence of voracious fish, and on a lithograph entitled Predestination (1951) these two independent figures go around in a Mobius-like (A. F. Mobius (1790–1868), German mathematician) strip in such a way that the innocent birds are devoured by the ferocious fish. These very same figures are depicted on a simple

---

[*] I am deeply saddened by the passing away of S. J. Gould on May 20th, 2002. He gave me thousands of hours of reading pleasure.

plate (**Figure 3.1**, reproducing the plate 29 from the book *Fantasy & Symmetry. The Periodic Drawings of M.C. Escher*),[5] to illustrate the method that the artist used. Two images with the same basic silhouette fill the plane.

Figure 3.1. Plate 29 from reference.[5] Two different motifs crossing the surface of the plane. A "ferocious-looking" fish and an "innocent" bird. Courtesy of IUCr.

Believe it or not, in crystallography there are two concepts that I have quite often connected to a counterpoint in three dimensions, a geometrical interweaving of two ways of describing the same three dimensional space, forming a spatial counterpoint. The person responsible for introducing this piece of geometrical wizardry was Paul Peter Ewald (1888–1985), another of the group of physicists that was involved in the original diffraction experiment described earlier. He was a student of Arnold Sommerfeld (1868–1951)—of atomic fame—and at the time of the first diffraction experiment described in Chapter 2, he was trying to develop a theory of the diffraction of light by crystals as a dissertation topic. The beauty of the concept is that by using his novel geometrical artifice, one can immediately interpret the complex diffraction patterns obtained by any diffraction experiment.

The concept is simple. The three-dimensional space that contains the crystal motif with the corresponding atoms is referred to as "real space" and the geometrical lattice that describes this space is the "real lattice," or x,y,z space. Recalling the diagram of the sheet of stamps from the first chapter as a two-dimensional crystal (now on **Plate 3.1**), one can see how the motif "Love" is repeated in a rectangular lattice of different axes, one horizontal (along **a**, shorter of the two) and one vertical (along **b**). The dots that mark the edges of the stamps (their perforations) form the lattice, within which the motif is embedded. As illustrated before, within that lattice one can join equivalent sets of points in the lattice resulting in parallel planes with equal separation. Also it was mentioned that the common direction (i.e., a vector) and the identical distance between them could describe a set of parallel planes. For mathematical convenience a perpendicular (or *normal*, shown as thin lines with small arrowheads) better describes the common direction of the set of planes. An inverse, "*reciprocal lattice*," can be constructed so that the direction of the planes is still described by the normal to the original planes, but now the spacing between them is related to the "inverse" of the constant inter-planar distance (look at the large dark spots regularly spaced in **Plate 3.1**). The points in this reciprocal lattice are labeled by three integers (for example: 200, 220, and 140, etc.; The point 000 represents the origin of both lattices). Mathematically, it can be proven that this "inverted lattice" has the same symmetry as the direct lattice and so the structural properties of the original lattice have not changed by this transformation even though the lengths have changed.

The construction principle has been illustrated in **Plate 3.1** where the background is our familiar two-dimensional crystal. Overlaying the crystal image, a series of planes of different colors have been drawn following the conventional crystallographic indices h,k,l. In addition, a portion of the reciprocal lattice, corresponding to the real lattice of the two-dimensional crystal, has also been constructed: notice the conspicuous dark circles regularly arranged and labeled with three integers (no commas). Observe how the grid points of the reciprocal lattice fall on to the thin lines perpendicular to the color-coded planes of the real lattice. Can you appreciate the two geometrical descriptions of the crystal (i.e., colored planes and ordered spots) overlaying the image simultaneously and forming an amazing geometrical tapestry?

The beauty of the construction is that by using this concept it is very easy to predict and analyze the diffraction patterns resulting from the irradiation of crystals by X-rays. The geometry of the Ewald construction is such that a set of planes (i.e., a reciprocal lattice point) will appear on the diffraction pattern as a spot if, *and only if*, the corresponding vector lies on the surface of a sphere of radius $1/\lambda$, where λ is the value of the wavelength or color of the radiation used. The power of this interpretation has been well-recognized in the crystallographic community and this imaginary sphere is named Ewald's sphere or the "sphere of reflection." It is acknowledged to be one of the most important concepts in the theory of diffraction. This geometrical construction guarantees that the set of planes with indices h,k,l on the surface of the sphere comply with Bragg's Law (**Figure 3.2**). In a pictorial way, the diffraction pattern is a magnified representation of the reciprocal lattice as its points intersect Ewald's sphere. As the real lattice containing the atoms within the crystal moves (typically it is allowed to rotate about a spindle axis by the instrumentation, **Figure 2.2**), the imaginary reciprocal lattice also rotates like a twin sister.* Reflections appear on the diffraction pattern when the reciprocal lattice points cross the surface of Ewald's sphere. Only then do they satisfy Bragg's law. You can check the simple algebra if you wish. This geometrical construction is shown in **Figure 3.2**.

A less abstract explanation of the same idea can be presented in the ensuing two paragraphs. Take a rubber band or an elastic string. Use a pin to fix one extreme on a flat piece of paper. The variable length of the elastic material can be used to mimic the variable wavelength of the X-ray radiation impinging upon the sample. Since the radius of Ewald's sphere is $1/\lambda$, we would stretch the elastic material outwards for X-rays of short wavelengths, and shorten it for longer wave-

---

* I am using "sister," because in Spanish "lattice" is feminine while "space," as in "real" or "reciprocal" is masculine—no offense intended.

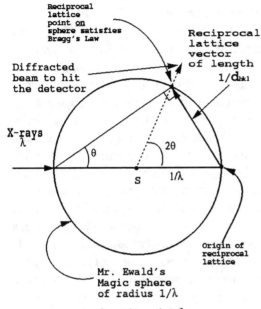

Reciprocal lattice point on sphere satisfies Bragg's Law

Diffracted beam to hit the detector

X-rays λ

Reciprocal lattice vector of length $1/d_{hkl}$

θ

2θ

s

$1/\lambda$

Origin of reciprocal lattice

Mr. Ewald's Magic sphere of radius $1/\lambda$

$\sin\theta = 1/d_{hkl} \,/\, 2/\lambda$; or
Bragg's Law: $2\, d_{hkl} \sin\theta = n\,\lambda$

Figure 3.2. The Ewald construction of the reciprocal lattice. Because of the geometrical construction, a point on the surface of the sphere automatically satisfies Bragg's Law, $\sin\theta = (1/d)/(2/\lambda)$, or $2d \times \sin\theta = \lambda$. Hint: Let me remind you from elementary trigonometry that the sine of angle θ is equal to the opposite side $(1/d)$ over the hypotenuse (diameter of the sphere, $(2/\lambda)$).

lengths (i.e., inverse relationship). One can think of the reciprocal lattice points that make up a crystal (the dark spots in **Plate 3.1**) as equally spaced beads on a string, or as equally spaced knots along a continuous rope. Each knot or bead represents a set of planes in the crystal with a specific orientation in three dimensions and a unique distance between them.

Using these two tangible objects—an elastic string for the radius of the sphere, and a string decorated with beads glued to the thread at equal distances to represent the points in the reciprocal lattice—it is easy to understand Ewald's construction. At a certain distance from the fixed origin of the elastic string (origin of the sphere), fix also one side of the bead-containing thread. The geometrical description of the earlier paragraph can be reworded to say that a diffracted beam will emanate from the sample *if and only if* a bead on the beaded string coincides with a certain elongation of the elastic material, extended from the origin of the sphere. A combination of variable radii (i.e., multiple wavelengths) with rotation of the crystal in front of the X-ray beam, maximizes the chance of satisfying the diffraction condition multiple times, resulting in multiple reflections on the surface of the detector. This physical model may help as you read the more abstract explanation in the earlier paragraph again, and study **Figure 3.2**.

Let me emphasize one more point in relation to the real and reciprocal lattices. The inverse relation between the two lattices also implies that the diffraction patterns have to be examined "inside out." That is to say, the minute details of very short inter-atomic distances (i.e., very small inter-planar separations or large reciprocal distances) appear as weak spots at what is referred to as "high resolution" on the outside of the diffraction pattern, away from the origin. These weak spots in the outside of the diffraction pattern are very difficult to measure accurately. Conversely, the gross features of the molecules in the crystal, their size and overall dimensions (i.e., large molecule-to-molecule distance or short reciprocal distances), result in very strong spots, close to the origin at "low resolution." The more reflections you are able to record from the crystal (both high and low resolution), the more detail your final molecular image will have. This is referred to as the "completeness" of the diffraction data.

You may wonder why is it that these simple concepts are not very popular even among crystallography students, and that the name of Prof. Ewald is not better known by people outside of the crystallographic community. There might be many reasons for that but I am going to suggest that part of the problem may be an underlying bias existing possibly in the neurology of the human brain, but certainly in our culture. A cultural prejudice between the terms "right handed" and "left-handed" with the related linguistic terms "dexterous" and "sinister" has been discussed before by S. J. Gould several times (see, for example, reference).[6] I would like to suggest that a similar neurological limitation and cultural animosity exists between the terms "direct" and "inverse" in our common language, making it more difficult for people to understand inverse relationships. In common language, a "direct" person is somebody straightforward, simple, and easy to understand; while "invert" or "inverted" has many connotations of abnormal, reversed, or twisted. We can clearly understand that if a variable increases by a certain amount, a related phenomenon will also grow by a proportional amount. However, we seem to have difficulty visualizing that if a quantity increases, a related quantity or variable might do just the opposite. Probably unknowingly, the introduction of the term "reciprocal" was an early public relations attempt to improve the image of Prof. Ewald's "inverse lattice" in the field. Mathematically, they are almost equivalent terms, but "reciprocal" sounds more intriguing. Both inverse and reciprocal suggest that if you repeat the process you return back to the original lattice, which is true.

Even so, I think that even students of crystallography are taken aback by the drawings and algebraic equations defining the "reciprocal lattice." In spite of its beauty and importance, the reciprocal lattice seems to be the "ferocious-looking" fish in the mathematical theory of crystallography. As a counter part, its associated sister is the smiling, sweet-looking bird that most people connect with the "direct" or "real" lattice. To study their atoms, molecules and structures, crystallographers use direct space labeled by three-dimensional vectors of components x, y, z; reciprocal space is their constant reference to analyze the diffraction patterns visually and mathematically, labeled by vectors in reciprocal space with three components h, k, l. The two geometrical constructions harmoniously interlace in a unique spatial counterpoint. Crystallographers have learned to juggle the two concepts simultaneously and consider one or the other depending on the context. In the magnificent tapestry that constitutes the theory and practice of crystallography, the concepts of real and reciprocal lattice are the subjacent warp and weft that support the equations, relationships and applications used and developed by crystallographers all over the world.

4

The Power of Waves: Fourier Analysis of Waves

Although conceptually very similar to unraveling the structure of common salt (i.e., NaCl, just two atoms), the problem of finding the atomic positions of more complex molecules could not be done by trial and error. Such methods could not possibly work for atomic structures containing hundreds or even thousands of independent atoms. New methods had to be devised to be able to uniquely place in space the minute balls that chemists interpret as atoms. The ingenious mathematical procedures that were developed form the core of theoretical crystallography, and are probably responsible for the dreadful reputation that crystallography has among the laity. It is considered to be hermetic, incomprehensible and even inscrutable. Please, do not panic. I have discussed the explanation that follows with many non-scientists friends, and with our teenage daughter Inés. They all found that it is easy to follow, imaginative, and even enlightening.

Imagine yourself in a concert hall, listening to a concert by a world famous symphonic orchestra. I am particularly fond of the Chicago Symphony Orchestra (CSO, Daniel Barenboim, music director). As much as you enjoy the sound, being able to see the spatial arrangement of the ensemble is also part of the experience. You can see the conductor surrounded by the first and second violins, violas, cellos, basses; winds and woodwinds behind and still further away the brass and the percussion instruments. All are in an organized group, complementing the harmonious sound coming from the entire orchestra. Close your eyes for a moment and ask yourself the following question: Without knowing where the different instruments are, could I reconstruct their respective positions, shapes and motions using just the overall sound coming out of the stage?

I must introduce a few physical notions before answering that question. The motion of the bow on the different strings, the wind blown into the wind instruments and the motion of the arms and hands of the percussionists all provide the energy that generates the sound waves emanating from the instruments. All of these waves combined reach our ears located at a certain level, row and seat of the concert hall. Imagine you could obtain similar (although weaker) sound waves by energizing the instruments with a very strong, intense and multicolored wind coming from the back of the stage. That being the case, you could politely ask all musicians to leave their instruments at their specific locations and, including the conductor, walk off of the stage for a few hours. You could amplify the signal by imaging many identical copies in a crystal-like arrangement. Let us assume that the stage can rotate, as in many modern auditoriums.

I would like you to capture this image and bring it into the realm of the atomic and molecular microcosm. Atoms are contained within a crystal. They are held in place by chemical bonds within each molecule and by weaker forces between the different molecules. For all practical purposes, a crystal is an infinite block of repeating units, and each "brick" or unit (the unit cell) contains a cluster of atoms that by displacement in three directions form the crystal (**Plates 1.1 and 1.2**). These displacements are called translations in the mathematical jargon of crystallography. Take a large piece of this block (containing several thousand "brick" units) and place it on the rotating orchestra stage, molecules taking the place of the musical instruments. Since a common model for many atomic phenomena in physics is to assume that atoms are minute oscillators that produce waves of a certain wavelength (or frequency), I do not think it is such a bad analogy. Let us leave aside the difference between sound and electromagnetic waves. Replacing the strong, intense wind current by a directed beam of X-rays will complete the analogy of reconstructing the position and identity of the musical instruments on the stage with the determination of the atomic structure using the diffraction (or scattering) of X-rays by crystals (**Figure 4.1**). More discussion on this topic will be found in Chapter 9 entitled *"The Combs of the Wind."*

Can we then reconstruct the positions, shapes and motions of the musical instruments (atoms) on the stage (crystal), using only the waves emanating from this imaginary experiment? Amazingly enough, at the atomic level the answer is yes. The work of crystallographers around the world involves the determination of the three-dimensional arrangement of atoms using the information obtained from the diffraction of X-rays by crystals. The tortuous path that allowed humans to achieve that feat is connected with the birth of Jean Baptiste Joseph Fourier on March 21, 1768, in the ancient French town of Auxerre. The city had escaped many of the major calamities derived from the plague and invasion by the Normans, and by 1751 was a prosperous town with a very high ecclesiastical status. This factor was probably not ignored by Fourier's father, a master tailor, when he decided to settle in Auxerre on the banks of the river Yonne, a tributary of the Seine.

Not unlike Abbot Haüy, J. B. Joseph Fourier began his education in the church, first at the abbey of St. Germain in his own hometown, and later in the beautiful and serene romanesque abbey of St. Benoît-Sur-Loire, Burgundy. This architectural jewel is located near one of the many trails that in the Middle-Ages connected northern Europe with the two routes across the Pyrénées to the tomb of the apostle St. James in Galicia, northwestern Spain. In medieval times, the flow of people and ideas along these walkways facilitated the dissemination of knowledge in the intellectual darkness of the era. As a small homage to the person who developed the mathematical foundations of modern crystallography, a visit to this medieval monastery should be a required pilgrimage for crystallographers old and young.

Also like Abbot Haüy, Fourier was subject to the vortices of the French revolution and during the last of the bloodiest episodes he was nearly put to death in the guillotine. Nevertheless, calmer times

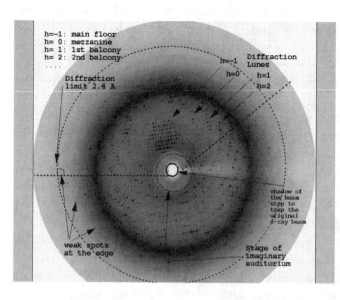

Figure 4.1. Illustrates an example of a diffraction pattern from a protein crystal when exposed to X-rays from a standard laboratory source, emitted by a copper target bombarded by fast-moving electrons. After filtering, the resulting X-rays are of essentially only one wavelength: λ is equal to 1.5418 Å.

Dimensions of the unit cell of this crystal are approximately **a**=90, **b**=90, **c**=105 Å. The pattern has been annotated to illustrate the suggested analogy between a diffraction pattern and a concert audience. The circle full of spots in the center is supposed to mimic the main floor of the concert hall. All the reflections in this circle will have as a "level" (first) coordinate −1. Within this circle, individual reflections will have two additional coordinates ("row" and "seat" or k and l). Similarly for the upper levels the first coordinate will be 0, 1, 2 for the "mezzanine," 1st and 2nd balcony respectively. The clear circular area in the middle is the shadow of the catcher that traps the original X-ray beam, and is labeled as the stage of the imaginary auditorium. The high-resolution data (atomic details) are the weak spots on the outside of the diffraction pattern; gross features of the molecule (i.e., size and shape) are related to the spots near the beam catcher (low resolution). The dashed circle in the outside indicates the limit of the diffraction pattern indicated as 2.4 Ångstroms (2.4 Å). Beyond this demarcation, the reflections are too weak to be measured. The crystallographic data extracted from this pattern consist of a list of reflections labeled by h, k, l (their reciprocal space coordinates) and their associated intensities as measured by the darkness of the spots on the pattern. A data set may consist of 30–90 frames similar to this one, recorded as the crystal rotates in front of the X-ray beam. Diffraction patterns produced by X-rays containing a mixture of wavelengths (or colors) contain many more reflections and are more difficult to interpret; they are referred to as Laue diffraction patterns for obvious reasons.

allowed him to attend the École Polytechnique and he was commissioned by Napoleon to take a prominent role in his campaign in Egypt as permanent secretary of the Cairo Institute. Following his return to France, he was appointed prefect of the department of Isère by Napoleon and worked for many productive years in Grenoble, the capital.

I decided to include these brief biographical notes to highlight the fact that after his return from Egypt to live in colder climates Fourier was very sensitive to cold.[1] This might have prompted his interest in the physical problem of heat propagation and heat losses. After several years of work, his mathematical treatment of this physical problem was published in 1822 (*Théorie Analytique de la Chaleur*; English translation *The Analytical Theory of Heat*, 1878) and had a tremendous influence in both physics and mathematics. Within this volume, Fourier used the idea that any mathematical function could be expressed as a "linear combination" (i.e., a simple combination of sums and products without squared factors) of waves using functions that contain $\sin(x)$ and $\cos(x)$ (see **Glossary**). This can be paraphrased in non-technical terms to say that we can sum simple wavy patterns together to produce essentially any shape.

Textbooks illustrate this principle in many different ways (see, for example, appendix 1 in reference),[2] and you can also simulate this calculation very easily on a graphing calculator. Using a musical analogy you will understand it immediately. Think of each of the piano keys as producers of a single wave corresponding to the frequency of the note (i.e., A natural is 440 oscillations/second or 440 Hertz). The combination (playing) of two or more different notes at the same time produces a more complex sound, which can be thought of as the "combination" or addition of two waves. When you combine two (or more) notes, one of the variables is how hard you press one note in relation to the other(s); this can be related to the factors or coefficients used in the combination. As long as you stay in a gentle regime, with no term "overpowering" the others, you can assume that you are doing a "linear combination." In the language of sound waves, Fourier's theorem meant that you could approximately break down a complex sound into its individual components by combining a certain number of basic notes with the appropriate intensities for each. Obviously, the more basic notes or "terms" you use the most faithful the representation of the original sound.* Conversely, it will take many terms to reproduce the complex sound of a symphonic orchestra.

How can we use this mathematical insight to solve the problem of finding the distribution of musical instruments (atoms and molecules) on the stage (crystal)? The important thing to emphasize is that you need to collect as much information as possible (this is further developed in a forthcoming Chapter 7 "*Ode to Data*"). Suppose that after the concert, or after a few hours of wind blowing on the stage, you ask to each and everyone of the people attending the concert two questions: i) where were you seated?; level, row, and seat please; ii) how much did you hear?; a number regarding the intensity of the sound waves reaching that particular position. These are the critical pieces of information that you need to solve the problem and you should strive to get data from everybody, otherwise your "diffraction pattern" will be incomplete.

Every spot (seat on the concert hall) on the diffraction pattern of a crystal (stage) is labeled in space with three integer numbers: h (level); k (row); and l (seat) (**Figure 4.1**). In addition, for each spot, you can measure how dark or intense it is; this value corresponds to the intensity of the sound waves reaching a position with coordinates: level, row, and seat. The mathematical theory developed by Fourier, and used much later to study the diffraction of X-rays by crystals, permits you to sum the contribution of each sound wave heard by each member of the audience sitting at position (level, row, seat) with all the others to give you a "shape" function. This "shape" function can be interpreted as a representation (almost like a shadow, cast or silhouette) of the original musical instruments (atoms) on the stage (crys-

---

* This analogy is probably not new. It occurred to me after discussions with my friend and colleague, crystallographer Eva Pébay-Péyroula from the University J. B. J. Fourier in Grenoble, France.

tal). Good old J. B. J. Fourier could not have imagined that we could use the ideas that he developed to explain heat losses to obtain the three-dimensional structures of molecules. Such is the nature of science. This connection was suggested by W. H. Bragg in 1915 and later demonstrated by his son W. L. Bragg in several examples in 1929.[3]

Waves, particularly ocean waves and surf, can evoke many emotions in humans depending upon the season of the year, and seasons and events of our lives. Their unceasing motion, their periodic and unrelenting pounding, each one different from the previous or from the one that just passed, speak to something deep within us. I always marvel at the fact that J. B. J. Fourier, in the dark, humid and cold winters of Grenoble, was able to harness the analytical power of the waves to allow us to unravel the atomic structures of crystals.

III

symmetry and properties of protein crystals

5

Can Crystals Cry?:
Hard and Soft Crystals

The crystals of the gemstones that are so important for the greed and comfort of the human spirit are hard, durable and especially rare. They can be polished, cut and faceted so that their value, appearance and glamour are enhanced from transaction to transaction: diamonds, rubies, sapphires, emeralds, aquamarines, opals, topazes and others. Nonetheless, crystals come in various sizes, shapes, and textures with many different physical properties. The external characteristics of a crystal are a reflection of the physico-chemical properties of the atoms and molecules that form it.

—Makes sense. But do they cry?

Let me continue.

Normally, the component molecules of a crystal are arranged in the geometrical lattice in an efficient and economical way. Chemical as well as spatial factors come into play to determine the minimum building block or motif (unit cell) of the crystals. This basic block, in turn, influences the external morphology of the crystals. I'll try to use simple geometrical arguments to illustrate these general principles.

—Can they cry?

I need a bit more background.

The structures of the simplest solids studied initially by W. L. Bragg (common salt—Sodium or Natrium Chloride—NaCl) were made up of spherical ions (Na$^+$ and Cl$^-$). Spheres are highly symmetrical objects because they can be rotated in any direction and by any amount and they still display the exact same appearance. As a rule of thumb—there are exceptions—the packing of very symmetrical objects results in highly symmetrical crystals, typically cubic or hexagonal. As the symmetry of the object to be enclosed in a cell gets lower so does the symmetry of the crystal containing it. For example, you can squeeze a spherical glass ball into the form of a light bulb. It is still highly symmetrical but only in a certain direction (the axis of symmetry). As you know very well, light bulbs are packed very neatly, head-to-tail in rectangular (or to be more precise, parallelepiped) boxes. Molecules containing a large number of atoms tend to have rather irregular shapes and therefore the number of ways to pack them within a crystal lattice becomes more limited. A simple way to illustrate this is to point out how pairs of shoes are stored inside their parallelepiped boxes, in a heel-to-sole manner. The boxes for shoes and boots or skates, are all parallelepipeds of different dimensions. In addition, large molecules like proteins tend to have certain "flexible parts" like the shoelaces of your footwear. The fourteen Bravais lattices discussed previously represent all of the unique boxes than can be used to build the large variety of crystals seen in nature.

—Seems reasonable, what about the crying of the crystals?

We are getting into it now.

Of particular interest for the protein crystallographers are molecules that contain atoms in the "organic domain" (Carbon, Oxygen, Nitrogen, Sulfur, and Hydrogen),[1] and that like shoes and gloves have a handedness. They cannot be superimposed on their mirror images in space. The simplest geometrical object with such a property can be obtained from a smooth sphere (central carbon) by attaching four toothpicks along the directions of the diagonals of an imaginary cube (see **Glossary** entry for **amino acid**). The four toothpicks have degraded the symmetry from that of a perfect sphere to that of a tetrahedron. If you now attach a different decoration to each of the four corners you will realize that there are two unique ways of doing it;

each one with a unique "chirality." This cluster of four objects is said to have a "handedness" (i.e., is chiral) in the same sense that our two hands are not superposable (see **Glossary**). They are mirror images of each other. The young Louis Pasteur (1822–1895, famous French chemist and microbiologist) was the first to make the connection between the handedness of crystals of certain organic molecules and some internal property of the corresponding molecules.[2,3]

—I do not see the point. Where is the crying?

Next paragraph please, and you will have an answer.

Proteins, and many other organic molecules, contain carbon atoms in their structure that have four different groups as neighbors; therefore they are chiral. In biological macromolecules, the majority of the times only one of the two possibilities is present. Protein crystals do cry.

—This is absurd. What do you mean?

I mean that because protein molecules are rather large and very irregular in shape, when packed in a crystal lattice they leave certain regions of the crystal empty and the space is occupied by solvent (typically water) forming what appears to be large tunnels or channels. Through these channels certain ions, heavy atoms (mercury, platinum), co-factors, substrates can diffuse into the proteins. By the same token, these ions (for example: calcium, Ca^{++}; and magnesium, Mg^{++}) and small molecules can leak out of the crystal into solution. In this sense, protein crystals do cry and sob. In some cases when these ions or atoms leech out of the crystal, the crystals crack (**Plate 5.1**).

—If they can cry, do they also age?

Yes, and very quickly. Two pioneers of protein crystallography Bernal and Crowfoot (J. D. Bernal, 1901–1971; Dorothy M. Crowfoot Hodgkin, 1910–1994) were the first to show in 1934 that protein crystals needed to be sealed in a capillary tube with a drop of "mother liquor" in order to diffract X-rays.[4] This simple but critical observation, was the beginning of protein crystallography.[5] Although the first diffraction patterns of protein crystals were extremely complex, they showed evidence of ordered atoms in all three direction of space, implying a well-defined atomic structure. If the internal liquid in the crystals dries out the crystals shrink and their internal order disappears.

—This is odd. What do you mean by "mother liquor"?

It is a term that protein crystallographers use for the solution in which the crystals have grown and developed. Typically, a solution of concentrated and homogeneous (i.e., pure with no contaminants) protein is prepared in a stabilizing solution or buffer. This is combined and equilibrated with a solution that contains a precipitating agent, normally salts, small organic molecules or organic polymers. Since most protein samples are very precious, all the experiments are done with very small drops in small plastic trays (**Figure 5.1**). After the crystals are grown, they are harvested in a "mother liquor" or stabilizing solution, which protects them from drying and is used to mount the crystals from the drops into the thin sealed capillaries (**Plate 5.1a–d**). Protein crystals are rarely very large. Typical dimensions are 0.2–0.5 mm on each side (approximately one to two hundredths of an inch; 1 inch equals 25.4 mm) (**Figure 5.2a–d**). Although crystals of this size can be seen with the naked eye, they are normally handled and manipulated under the microscope.

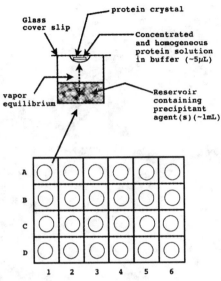

Figure 5.1. Schematic representation of a protein crystallization experiment. In the lower part of the diagram a two dimensional array represents the crystallization plate that is commonly used to grow protein crystals by the vapor equilibration method. Crystallization conditions are typically varied across the plate (numbers 1–6) and vertically A-D in the different wells. Important parameters for obtaining the most suitable crystals are the type and concentration of the precipitant agents as well as the pH of the protein and reservoir solutions. The size of the plate is approximately 15 × 10.5 centimeters (6 × 4 inches). This experimental setup allows for twenty-four conditions to be tested with about 50–100 μL (1 μL contains one millionth of a liter) of protein solution. In the upper diagram a single well is magnified to show the details of the experiment. A drop of a homogeneous and concentrated protein solution is layered upside down ("hanging drop") on a glass cover slip and sealed looking towards the bottom of the well where a reservoir solution containing approximately 1 mL (note the different scale of volumes: 1mL is equal to one thousandth of a liter) of buffer and precipitant solution had been deposited previously. Precipitants can be salts, small organic solvents such as ethanol and also large organic polymers like polyethylene glycol. Vapor equilibration takes place between the drop and the reservoir and, if the conditions are right, small protein crystals appear in the drop. These crystals can be examined under a low power microscope through the clear glass cover slips and the transparent plates (**Fig. 5.2a–d**).

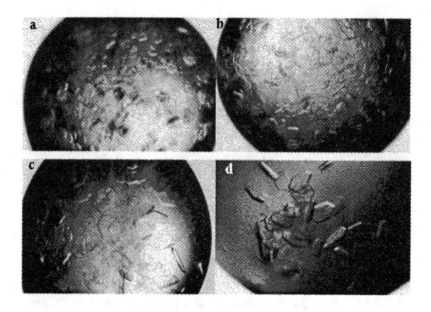

Figure 5.2a–d. Successive improvements in the quality of protein crystals. The panels illustrate the relative improvement in the quality of the protein crystals, as the crystallization conditions are fine-tuned. (**a**) is the result of the first screening for crystallization conditions using a standard set of buffers and precipitants. Finer exploration of the original parameters (pH and precipitant concentration) yielded crystals larger in size and with better defined edges and faces (**b, c**). The image (**c**) shows the typical, hexagonally-shaped, plates of a native protein called ErmC' (approximate dimensions $0.200 \times 0.200 \times 0.005$ mm^3). The well-defined and thicker hexagonal crystals of the last panel (**d**) result when in the same protein, the amino acid methionine is replaced by the selenium-containing methionine (see Chapter 9 and **Glossary**). The external appearance of the crystals is somewhat different but their internal structure is the same (i.e., isomorphous crystals; cell constants a=b=147.4, c=58.1 Å). Details of the crystallization conditions and of the atomic structure within these crystals can be found in reference.[7] ErmC' is a protein that is part of the biochemical machinery used by bacteria to develop resistance against antibiotics. Diagrams prepared by the author with the assistance of Jeff Frye from Abbott Laboratories' Creative Network.

—If the protein crystals cry and age, they must also die? Do they?

Yes, they do. When mounted in capillaries, after long exposures to the X-rays crystals cease to diffract and, in a way of speaking, "die." Nonetheless, in the last few years a new mounting method is being used, which extends the lifetime of crystals in front of the X-ray beam. Protein crystals are currently flash-cooled at very low temperatures (100–150 degrees Kelvin, approximately 170–120 degrees Centigrade below the freezing point of water) using liquid nitrogen, and for all practical purposes they become "immortal." They continue to diffract for an almost indefinite time.

Crystals of biological polymers, and in particular crystals of proteins and viruses, are probably the softest and most fragile of the crystals in the natural world. They are born, they grow, they cry, they age and they die. These characteristics are a reflection of the properties of the molecules that make life possible and stand in contrast to the rocks and minerals of the inanimate world.

6

Remembrances of the Alhambra: Symmetry in Space and Time

"Dale limosna mujer, que no hay mayor pena que la de ser ciego en Granada"
Miguel de Cervantes, author of *Don Quijote*

To a woman in front of a beggar at the entrance of a church,
"Woman, do not deny him the alms, because there is no greater sorrow than being blind in Granada."
(translation by the author)

Everybody responds to symmetrical objects or buildings through an intuitive sense of equilibrium or balance of proportions. Symmetry is indeed prevalent in our common artifacts and buildings; yet, the abstract description of symmetry is normally hidden behind its harmonious appearance. I would like to invite you to discover the classic book *Symmetry* by Hermann Weyl.[1] Beginning with the common notion that symmetry equals harmony of proportions (*syn-metron*, probably the original etymology of the word), he takes the reader in a delightful tour of the different types of symmetry in art and science, to culminate in a discussion of crystal symmetry. As described before, the unique character of crystal symmetry is related to the repetition in three dimensions of a certain atomic or molecular building block. However, there is a little bit more behind this notion and I would like to explore it with you in the next few paragraphs.

The idea of a symmetrical framework underlying the description of our physical world has had a long tradition in western science from the Greeks to modern times. In crystallography, symmetry is the essential concept upon which every aspect of the field starts, develops and expands. There are several types of symmetry that need to be understood. The fundamental symmetry that defines a crystal is the translation along the main directions of the crystal of a box-like entity called the unit cell. In addition, the unit cell content itself might also be generated by the repetition of a simpler motif, using rotations and displacements within the unit box. This is quite common in the patterns of fabrics or wallpaper. In the example of the sheet of stamps (**Plate 1.1**), it is apparent that there is no repetition within each stamp. However, one can envision that other stamps displaying more symmetrical images could indeed contain a smaller fragment that could be duplicated several times to generate the entire image. For instance, one can generate an entire hexagon using only a triangular wedge repeated six times by a rotation of sixty degrees (i.e., rotational symmetry). Duplication by rotation is common in many decorations of plates, bowls and dinnerware since the purpose is to fill a circular dish. **Plate 6.1** illustrates 3- and 6-fold symmetrical pattern. The most striking examples of rotational symmetry in three dimensional space are the five platonic solids: tetrahedron (four triangular faces); cube (six faces); octahedron (eight faces); dodecahedron (twelve faces); and icosahedron (twenty faces). As you will read later, the icosahedron is the geometrical scaffold for the molecular structure of many viruses.

The simplest form of repetition is displacement or shifting of the original image from one place to another without distorting the image (examine **Plate 6.1** carefully). From the earlier example of a unit cell containing two neatly packed light bulbs, one can visualize how to generate the pair of light bulbs by starting with only one and produce the other by an appropriate rotation (180°) and a shift (rotation and translation symmetry). At a more complex level and in three dimensions, one can construct a spiral staircase by rotating the first step about the vertical axis by different amounts and shifting the new positions upward by the appropriate distances (i.e., screw symmetry operations). All of these types of symmetry are found hidden below the regularities of crystals. All these types of symmetry have been superbly illustrated in the book *Symmetry—A Unifying Concept* by Istvan and Magdolna Hargittai.[2]

In crystallographic terminology, the smallest building block that can be used to generate the contents of the unit cell is referred to as the "asymmetric unit" of the crystal. The movements or actions needed to fill the unit cell with copies of the asymmetric unit are called symmetry operations. What is inside the asymmetric unit of a crystal is like a seed that multiplies by symmetry operations to generate the unit cell. The content of the unit cell then propagates throughout the crystal by displacement along the axes of the crystal. In spite of its name, the contents of the asymmetric unit (a-symmetric: no symmetry) can have symmetry by itself and this has important consequences for the study of large biological structures such as viruses. This symmetry not generated by the crystal is appropriately called non-crystallographic symmetry and you can read more on this subtle point in later chapters (Chapters 10 and 12).

In the graphic arts, nobody has grasped the concept, scope and intricacies of crystallographic symmetry better than the Dutch graphic artist Maurits Cornelis Escher (1898–1972). Because of this, crystallographers are particularly fond of him. I am no exception. Crystallographic symmetry is ever present in his woodcuts and engravings. This fact is well known in the art world, and there is a monograph focusing on this aspect of his work entitled *Fantasy and Symmetry: The Periodic Drawings of M.C. Escher*,[3] which you can use to gain a better understanding of the richness of crystal symmetry. However, I expect to entice your curiosity by bringing up the issue of the origin, inspiration, and motivation of the symmetrical elements found in the graphic art of this master craftsman. As we explore the places that he visited in his travels through Europe, we will realize that symmetry is also present in the air that surrounds us, generated by gently blowing winds or human hands that anonymously enrich our lives with all kinds of symmetrical sound patterns.

M. C. Escher was born on June 17, 1898, in Leeuwarden, Northern Netherlands. It was very soon evident that he liked to draw and was encouraged by his early teacher, Mr. F. W. van der Haagen, to make prints. On this father's advice Escher went to Haarlem to study architecture. One of the faculty members there, the Dutch artist Samuel Jessurun de Mesquita (note the Arab surname), advised him to leave architecture and concentrate on the graphic arts. Escher had already realized that he did not like architecture very much and followed his mentor's advice wholeheartedly. He continued his education under the guidance of Jessurun de Mesquita from 1919 to 1922, and afterward traveled frequently to Southern Europe, particularly Italy and Spain.

In the spring of 1922, Escher went to Italy, and in the autumn of the same year he made a brief trip to Spain. This trip to Spain made a tremendous impression on him. From 1923 to 1935 he settled in Italy, making frequent trips to small villages in Northern Italy, Sicily, and Corsica, always taking notes and making sketches of motifs of interest which he later used for prints during the winter. The political situation and the rise of fascism in Italy made his life there less pleasant, and for a short time in 1935 he moved to Switzerland. From May through the end of June in 1936, he made his longest study trip along the coast to Italy and Spain. On this voyage, he made detailed copies of the Moorish mosaics in what to me are the two most outstanding monuments of Arab architecture in western Europe: the Palace of the Alhambra in Granada, and the Mezquita Cathedral at Córdoba, both in Southern Spain. After a short residence in Ukkel (near Brussels), Escher returned to the Netherlands in 1941 to settle in Baarn. He passed away in Laren on March 27, 1972.

After 1937 M. C. Escher became rather sedentary and traveled only for family reasons. His trips did not have any further influence on the material of his work. During a trip to Canada in 1964 to visit his oldest son, he was scheduled to visit the United States and give a series of lectures discussing his methods and techniques. Unfortunately, all the engagements were canceled due to illness. He was admitted to Saint Michael's Hospital in Toronto to undergo emergency surgery and upon his return to the Netherlands his health continue to deteriorate. However, the intact, fully prepared, text of these lectures with the corresponding illustrations, has been published in English,[4] and, through them, the reader has a rare glimpse of the artist's techniques and "intentions."

Regarding his interest in the regular division of the plane he wrote:[5]

"I can't say how my interest in the regular division of planes originated and whether outside influences had a primary effect on me. My first intuitive step in that direction had already been taken as a student [...] in Haarlem. This was before I got to know the Moorish majolica mosaics in the Alhambra, which made a profound impression on me."

In the opening remarks of the first lecture he explains in more detail:[6]

"Many of the bright-colored, tile-covered walls and floors of the palaces of the Alhambra in Spain show us that the Moors were masters in the art of filling a plane with similar, interlocking figures, bordering on one another without gaps. [...].

What a pity that the religion of the Moors forbade them to make images! It seems to me that they sometimes came very close to the development of their elements into more significant figures than the abstract geometric shapes that they created. No Moorish artist has, [...] ever dared (or didn't he hit on the idea) to use as building components concrete, recognizable figures borrowed from nature, such as fishes, birds, reptiles, or human beings. This is hardly believable, for *recognizability* is so important to *me* that I never could do without it.

Another important question is *color contrast*. It has always been self-evident for the Moors to compose their tile-scenes with pieces of majolica in contrasting colors. Likewise, I have never hesitated myself to use color contrast as a means of visually separating my adjacent pattern components."

So, what is the Alhambra that left such a lasting impression on M. C. Escher? It is the last and most exquisite outpost of Arab civilization in Western Europe. The last dynasty of Moorish kings reigned there until the incipient nation of Spain, under Queen Isabella of Castille and King Ferdinand of Aragon, conquered it in 1492. The fall of Granada under Christian rule opened the way for the adventure of Columbus and the New World. Atop the most gentle ranges of the Sierra Nevada, the Alhambra looks like "a square, brutal fortress" that does not reveal (nor even hint) at the delicacy of the forms buried in its interior.[7] Inside, the Alhambra is a palace of exquisite, intricate, and delicate beauty. It is a labyrinth of columns, corridors, courts, chambers, and windows—full of geometric and sensuous forms, quivering with the music of water murmuring in the courtyards. I speak from experience when I say that even crystallographers will get intoxicated and dizzy by the ever-present symmetry in the colorful majolica tiles on the walls and floors, carved on the plaster in the ceilings, inscribed on the supporting wooden beams, and surrounding the secluded windows. I can only recommend that you visit it.

Plate 6.1 depicts an overlaying of two images. In color, you can see the intricate symmetry and design of the Moorish artists in southern Spain as it is found the Alhambra. Geometrical figures fill the plane with symmetries involving rotations and translations, combined in kaleidoscopic and muticolored designs. Overlaying in black and white, you can see the first page of the score of a musical composition for the classical guitar. As you can appreciate, the basic pattern of every measure of the musical page, and indeed every measure of the entire piece, is essentially identical. Both the treble and the base line have the same structure in the horizontal direction; only the individual notes change.

As much as music can evoke places and sensations, I will venture that there is a very special composition of music for the classical guitar that does succeed in evoking in my brain the interior of the Alhambra. This is precisely the piece from which I have reproduced the first page of the score in **Plate 6.1**. Musically, it is a very modest piece. In fact, it was intended more as an *étude* than a full composition and is entitled very appropriately *Recuerdos de la Alhambra*. It was written by Francisco Tárrega (1852–1909), a Spanish guitarist and composer who played a very important role in developing the music and technique for the classical guitar in this century. In doing so, he was instrumental in transporting this delicate instrument from the *tavernas* and *cafés* of 19th century Spain to the concert halls of the world. Classical guitar music is now enjoyed by a multitude of people on the recordings of Andrés Segovia, Narciso Yepes, Julian Bream, Cristopher Parkening, John Williams, Sharon Isbin, and many others.

Being an *étude*, *Recuerdos* is not a very exhilarating piece. As I pointed out before, it is monotonous and repetitive, as symmetry is. Yet, it is a serene and beautiful piece that has become an indispensable composition in the repertoire of the romantic guitar. It attempts to exercise a technique of particular importance to the string instruments and particularly to the classical guitar, named "*tremolo*." It consists of the rapid reiteration of a note or group of notes, resulting in a "quivering" or "tremulant" effect. Not inappropriately, "*tremolo*," has the same root as "tremor" and "*tremuloides*." In its most gentle mode, the tremolo sounds like the quivering of the leaves of the quaking aspens (*Populus tremuloides*) ever present on the slopes of the Rocky Mountains in the Western United States. It is specifically this "quivering" sound which I associate with the murmur of running water of fountains and streams, in combination with the monotonous, symmetrical nature of the music, that my brain and my senses associate with the serenity and enchantment inside the walls of the Alhambra.

In our macroscopic world, we recognize and appreciate symmetry by a sense of balance and equilibrium in a static setting:, immobilized in our buildings, frozen in our artifacts or paralyzed in our sculptures and tile work. There is, however, an infinite variety of symmetrical forms in the world of sounds, and especially in bird songs and musical composition of which we are barely aware. In addition, the atomic domain is inundated with regularities and symmetries whose reflection we appreciate in natural objects such as plants, flowers, and jewels. This is the realm of the atoms and molecules that constitute our biological and inanimate world. Crystals, crystallography and X-rays are the key elements to unveil the intricate patterns and labyrinths of this domain.

7

Ode to Data: Uniqueness of Crystallographic Data

Our years of graduate study at the University of Texas at Austin were indeed happy ones. Both my wife and I were making progress in learning a new language. We were being exposed to a new culture, meeting interesting people, making new friends and experiencing a different way of doing things. The academic and scientific resources of the university appeared to be limitless. She was pursuing studies in foreign language education and I was finally able to follow my dream of pursuing graduate work in molecular biophysics, with a focus on protein crystallography. This was still a new and demanding undertaking. Our weekly routine was full of classes, homework, exams and our grocery shopping. On Fridays though, we would always go to the movies in one of the local theaters near "*Guadalupe*" Street with our friends Jaume and María Angels. After the movie, our weekend routine usually included a trip to the laboratory to develop an X-ray film from my crystallography experiments.

As I outlined before, I would have typically left a crystal being exposed for few hours to the X-rays and being rotated in front of a piece of film, enclosed in a sealed cassette as schematized in **Figure 2.2**. After the movie and some pastry and coffee, we would walk to the lab. I would take the cassette to the dark room adjacent to the lab; develop the film, fix it and then finally rinse it off with water. After this laborious process I would come out and announce to Vithorita, waiting outside, whether it was a good or bad exposure. I always invited her to look with me in the light box the good diffraction patterns, the ones much like the one presented before in **Figure 4.1**. She always wondered what I saw in those "little dots" on a square piece of film approximately 4½ inches on the side. I explained to her that those "little dots" were our data. Next, I would carefully prepare for the next twenty-four hour exposure that could collect—everything going well—up to eight films in a carrousel-like arrangement. A complete data set would typically consist of thirty to sixty films and might contain thirty to forty thousand unique "little dots." These data can be collected nowadays in less than an hour using intense X-ray sources and the new generation of detectors. In a way of speaking, those tiny dark dots orderly peppered on the surface of the detectors, flashing from frame to frame, are the response of crystalline matter to a concrete series of questions posed by the experimenters. This response is analyzed daily by thousands of crystallographers during experiments all over the world. You may ask, in which way is crystallographic data collection different from other data-gathering scientific activities? I will answer the question in detail but I'll put the question in a more general context first.

Our expanding knowledge of the material world around us is the result of a constant dialog between Nature and humankind. Except for special occasions, the first interlocutor is passive and lets the other observe its patterns and flows, broken only under special circumstances by magnificent displays of power, craftsmanship or beauty which are specially recorded by the keen eye of the observer. To paraphrase Albert Einstein (1879–1955), one of the icons of this unending dialog, one can say that in this ongoing conversation, "Nature is subtle but not malicious."* Quite often she speaks *sotto voce*, in a whis-

---

* "The Lord God is subtle, but malicious he is not," "*Raffiniert ist Der Herr Gott, aber boshaft ist Er nicht,*" in the German original. Quoted widely in various translations. Some say that Einstein meant that Nature hides her secrets by being subtle.[1]

per. Yet, she neither deceives nor confuses but just allows herself to be scrutinized and cross-examined in as many ways as the other participant can think possible.

In an attempt to simplify the complexity of this dialog, humankind began in the Renaissance to ask specific questions under controlled conditions, and names like Galileo Galilei (1564–1642) have been hailed as pioneers of this approach. An experiment is a very powerful way to attempt to get concrete and specific answers to questions from Nature. I say "attempt" because even under controlled conditions Nature may give ambiguous or puzzling answers. It is not her fault. When exploring the unknown, we often ask the wrong questions or formulate incisive questions within an inappropriate context. Our experiments may have faulty designs, or the results from the experiment(s) may fail to give us the "expected" answer within a certain hypothetical or idealized theoretical framework within our brains.

I should digress for a paragraph to make the following point about scientific data gathering. Any individual with a keen sense of observation and detail, committed to taking careful notes on any aspect of our environment, can participate in the dialog and add to our knowledge of the material world. Such a person does not have to be invested with special mathematical powers to change the way we view the world surrounding us. Keen observers have always been critical in our path towards scientific understanding. People such as Louis Pasteur (1822–1895) and Charles R. Darwin (1809–1882), both keen observers, have changed the way we see the world solely because they were masters at interpreting their attentive observations. In the domain of the behavioral sciences, we may learn more about ourselves by carefully observing the tribulations of groups of chimpanzees at the Gombe Stream Reserve on the shores of lake Tanganyika for forty years, as presented by Jane Goodall and others,[2-5] than by sequencing the entire human genome. How the unveiling of the human genome will expand the frontiers of understanding human behavior can only be a matter of conjecture.

In contrast, the physico-chemical sciences typically use in their experiments a pure (homogenous) sample containing only one or a few types of molecules in aqueous solution, normally at high concentrations. Biochemists and enzymologists are examples of scientists whose work involves analyzing in detail the interactions of a few types of molecules in solution. Their goal is to get answers to specific questions about the rates or speed of reactions, the effect of small activators (co-factors), and so on. Their work returns limited (but often very important) information regarding the behavior of such systems.

Crystallography and Nuclear Magnetic Resonance (NMR) are the most powerful techniques to obtain three-dimensional information. These methods also use homogenous molecular samples; they are no exception. I have presented briefly before how concentrated solutions of protein molecules can be crystallized (see Chapter 5 *"Can Crystals Cry?"*). For NMR, the sample is concentrated and subjected to a series of energy pulses normally directed towards the nucleus of hydrogen atoms (i.e., protons). As the energy is distributed near the surrounding atoms, NMR spectroscopists obtain clues as to which atoms are next to which other atoms in three-dimensional space. When they put all the pieces of the puzzle together, they obtain a three-dimensional structure of the molecule in their *cuvettes* with a limited amount of detail. Typically their data consist of several hundreds to a thousand data items.

The design of an experiment or set of experiments intended to determine the three-dimensional structure of a protein by diffraction of X-rays is not very different from the original experiment of Max von Laue and collaborators. First, there is an intense, well-collimated and focused source of X-rays, normally of only one "color" or wavelength (typically 1.5418 Å). Second, there is a crystalline sample inside a capillary, or frozen using a solution that prevents ice formation inside and outside the crystal. This crystal can be rotated under controlled conditions in front of the X-ray beam. Third, there are detection devices to record the different sections of the diffraction pattern. A schematic representation of this experimental set up has been presented before in **Figure 2.2**. It is the character, uniqueness, and in the end the completeness of the diffraction data as illustrated in **Figure 4.**1, which allow crystallographers to reconstruct in exquisite detail the atomic image of the molecule(s) inside the crystal. Rapid computer programs using algorithms based on the Fourier summations discussed in Chapter 4 (*"The Power of Waves"*) are commonly available to calculate the electron density (or "shape function") inside the unit cell of crystals.

As a reflection of the internal symmetry of the crystals, a diffraction pattern consists of thousands of data points arranged regularly in three-dimensional reciprocal space. The data points (reflections) extend from the center region, near the shadow of the catcher of the direct beam, to the outer regions of the surface of the detector. Circles can be drawn in a diffraction pattern enclosing reflections to different "resolutions," from low (inner part) to high resolution at the edges of the diffraction pattern. As the circles expand, a larger number of reflections will be included, resulting in "higher resolution" details in the map. The number of reflections collected on a crystallographic data set ranges from several tens of thousands (typical protein structures) to several millions for viruses. The amount of data is proportional to the volume of the unit cell and therefore to the size of the molecule, so that larger proteins typically give more data points. Moreover, because of the crystal symmetry, a reduced, averaged, set of unique reflections with internal agreement better than 7% or so are routinely obtained, which form the basis for the structure determination. Four electron density maps calculated at different resolutions are presented in **Plate 7.**1 superimposed with the atomic model obtained at the end of the structure determination process. Notice how the atomic nature of the polypeptide chain is more apparent in the highest resolution map (1 Å, lower right) than in the other three. Consequently, the positions of the atoms (colored balls: blue, nitrogen; red, oxygen; white, carbon) can be placed with more confidence and less error.

A point, let us say an atom, can be placed in a unique place in three-dimensional space by three coordinates (x,y,z), and a measure of its motions in space can be given by an additional parameter called the temperature factor (B). This means that we can describe an atom as seen by the X-rays using four variables. Therefore, a very accurate representation of a protein molecule containing 3,000 atoms can be given by obtaining values for 12,000 variables. A typical crystallographic data set for such a protein can contain well over 50,000 reflections at a resolution of 2.0 Å. Therefore, the ratio of observations to unknowns is approximately four. Nowadays, for some of the largest viruses being studied (for example, HK97, a bacteriophage containing double-stranded DNA, is 660 Å in diameter), a data set contains over 20 million reflections, with a reduced set of unique reflections having almost five million reflections to approximately 3 Å resolution. Combining these vast amounts of crystallographic data with the well-established rules of chemistry (inter-atomic distances and angles), protein crystallographers can determine the three-dimensional structure of most proteins and viruses to within a few tenths of an Ångstrom.

Due to the recent availability of synchrotron sources, the extent and quality of the data can be improved even further and structures of proteins with each individual atom resolved are becoming more and more common (**Plate 7.2a–b**). The details that can be observed at this resolution open unique opportunities for a better understanding of the function of biological molecules and has been described by some of the workers in the field as a "new universe."[6] This is what makes protein crystallography so unique and powerful. It provides the means for determining the atomic structures of many of the macromolecules that make life possible in exquisite detail. This is possible because of the enormous amount of data collected, and because of the quality and internal consistency of the individual data items.

In the dialog between Nature and Humankind there is a constant interplay between data gathering and inferences. Hypotheses or models are drawn from the data. A few items of data allow for several hypotheses to be consistent with those data points. As the quality of the data is improved and the amount of data extended, fewer models survive the filter and ideally we are left with only one model that can explain all the data. The redundancy and extent of the data obtained from protein crystals allow crystallographers to unveil a three-dimensional model of those protein molecules at the atomic level that satisfies all data. Using an artistic metaphor, protein crystallographers are like sculptors. The fruit of their labors is an atomic sculpture carved and chiseled with a level of finesse and detail consistent with the quality and extent of the data they collected. May this brief essay be a written celebration and a collective Ode to Data. Crystallographic data, a vast collection of "little dots," allows us to reconstruct in exquisite atomic detail the molecular entities that comprise our world, including ourselves.

IV

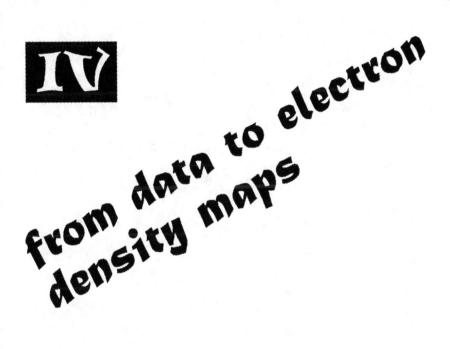

from data to electron
density maps

8

The Gordian Knot of Crystallography: The Phase Problem

I am sorry to disappoint you but as extensive as the data coming from crystallography are, they are only one part of the total. The detectors that crystallographers place in the back of their crystals can only record the magnitude of those X-ray waves at their h,k,l locations; they cannot measure their relative time positions. In the same manner, our audience in the concert hall can only hear the intensity of the sound at their assigned seat positions, not the relative separation of the sound waves. This is the "Gordian knot" of crystallography referred to as the "**phase problem.**" Unlike Alexander the Great, who chose to cut the knot with his sword, crystallographers have learned to untie this knot in several elegant ways, and in doing so they can obtain images of the atoms in the crystals by direct Fourier sums. Together we will patiently learn to untie the knot ourselves in this chapter.

The discussion on Chapter 4 on the superposition of waves was an oversimplification. I left out a very important detail that I want to present now. The terms in the Fourier sums necessary to obtain the shape function of the crystal have two parts. One part expresses the magnitude of the wave, and the other part expresses its relative position in relation to the other waves in the sum. In our simple sound analogy, the combination of two (or more) notes depends on the relative intensity of the two notes (their relative magnitudes) *and* on the difference in time of travel from the point at which the notes were played and the point at which the combined notes are heard. Notes played at different points on the stage will have different distances to travel to reach the seat at which we are enjoying the concert. This is why in the first example I specified that the two notes be played simultaneously. The combined sound of the two notes will be different if I play them simultaneously or with various time intervals. Note that the time intervals we are talking about here are in terms of the times for one cycle of the sound waves. At 440 Hz (A natural) this means time differences around 2 thousandths of a second (2 miliseconds), during which time the sound will travel about 2 ½ feet.

In mathematical terms, this added complication can be handled by saying that the terms in the Fourier combination or sum are complex numbers. Complex numbers were invented to get around the problem of obtaining the square root of a negative number, namely the square root of -1.[*] They consist of two parts, a real part and an imaginary part. They are typically represented in a system of axes where the horizontal (x) axis is the real part, and the vertical (y) axis corresponds to the imaginary one. An arrow (i.e., a vector) is also used to represent complex numbers, where the length of the arrow is the size (i.e., the modulus), and the angle it forms with the x-axis (counter-clockwise) is the phase angle, often just called the "phase." Think of it in terms of a hand in a wall clock. How long the hand is corresponds to the modulus; where it is in relation to a reference point (i.e., 12 noon) is the phase. For instance, if the hand is at three, then the phase would be $-90°$ (remember, clockwise is negative). Graphically, vectors are added by placing the tail of the second vector on the head of the first vector, without changing the phase angle of either one. The sum is an arrow drawn from the tail of the first vector to the head of the second vector. The real and imaginary parts of the resulting

---

[*] You can learn more about the history of imaginary numbers in the book.[1]

vector are obtained by adding separately the real and imaginary parts of the two elements of the sum. In this way, the combination of all the waves in the Fourier sums amounts to the sum of vectors, each one with its length and associated phase angle.

How can crystallographers calculate the phases of those measurements associated with the intensities of the reflections they record? By several methods, some experimental and some computational; in either case the methods are a triumph of ingenuity. The simple trial-and-error methods that W. L. Bragg devised in 1913 to obtain the structure of the simplest crystalline compounds (i.e., common salt, NaCl) could not be used to unravel the structure of more complex organic molecules such as table sugar, sucrose. Step by step the Braggs and their former students and colleagues extended the methods. They pioneered the use of Fourier summations to map the atoms of ever more complex molecules. The method reached a significant peak with the structure of the family of compounds named phthalo-cyanins by the Scottish crystallographer J. M. Robertson in the 1940s.

Independently, mathematical relationships between the values of the magnitudes of the different reflections of a diffraction pattern were developed for data obtained from small molecules (< 100 atoms). Many theoretical crystallographers worked on this problem and together they found ways of extracting reliable phases only from the magnitudes of the reflections making up the complete diffraction pattern. Initially, D. Harker and J. S. Kasper proposed a series of relations based on classical mathematical inequalities in the 1940s. W. Cochran, D. Sayre, P. Main, and many others investigated the power of probabilistic methods later (see, for example, reference).[2] All these theoretical developments reached a significant maturity in the formulations of H. A. Hauptman (b. 1917) and J. Karle (b. 1918). Their methods, and others derived from them, are now coded in ready-to-use computer programs referred to as **"direct methods."** With today's computers, these algorithms can map the atomic structure of organic molecules containing up to 200 atoms, using only the data from the diffraction pattern, within minutes of measuring the data.

For biological macromolecules like proteins and nucleic acids, each containing thousands of atoms, the road was also tortuous. In the end an elegant solution was found by Max F. Perutz (1914–2002) and colleagues, related to the method used by J. M. Robertson to solve smaller molecules. If you develop an interest in protein crystallography, you will learn that the method is called **M**ultiple **I**somorphous **R**eplacement (**MIR**) and you can read all the details of its application to the first protein structures solved, hemoglobin and myoglobin.[3]

The concept of **MIR** is easy to explain if you can flash back to the metaphor of the group of musical instruments on the stage and the audience in the concert hall with an intense wind blowing behind it (Chapter 4 *"The Power of Waves"*). Suppose that by some trick you could introduce into the group of musical instruments a very large one (i.e., a tuba) *without disturbing significantly* the relative positions and order of all the other instruments. This minimal disturbance would have left the crystal lattice untouched but altered the overall sound coming from the orchestra significantly. Assume that the people in the audience are unaware of this modification. They cannot see the additional instrument but can only "notice" that the overall sound that they hear at each seat in the concert hall is different.

Once again, you will need to record from everybody in the audience their position and the new values for the intensity of the sound. The mathematical analysis of the problem shows that you need to do this trick at least two different times (assuming you are dealing with perfect data). You can use the same heavy instrument (heavy atom) but it has to be at a different position. These two additional data sets combined with the original measurements from the unperturbed group of instruments (called native data) allow you to estimate the phase component of the Fourier terms. In order to minimize errors, you might like to keep the same people in the audience to carry out a third or even more experiments. It would definitely be a *very* long concert. The method is called "multiple" because you need more than two heavy atoms, and is called "isomorphous" because the crystal lattice is left practically unchanged. Once the phases of those individual observations have been estimated with reasonable precision, computers can calculate an electron density map very rapidly. I speak from heart when I say that even now, after almost thirty years in the profession, I feel a shiver up my spine just before an electron density map is displayed on the computer screen. In the past, the calculation of an electron density map represented the culmination of many years of work and whether the map was good (i.e., "interpretable" in terms of a protein chain) or bad (i.e., poor or noisy) was linked to the success of the project. You can appreciate how elegantly the multitude of waves combine at very high resolution to show individual atoms in **Plate 7.2a,b** or just a shadowy silhouette at very low resolution (5 Å) with improving tones in between (**Plate 7.1**). You will find other electron density maps along the way (**Figure 11.2**, 2.4 Å; and **Plate 15.1d**, 1.8 Å).

Prof. Max Perutz has written extensively about the early work on protein crystallography[4] and how the unveiling of the first three-dimensional structures of proteins had tremendous impact in physical chemistry, biology and biomedical sciences (see Chapter 19 "*Molecular Docking and the Broken Heart*"). All of these developments follow from that magic moment that he has described[5] when he compared one section of the diffraction pattern of two crystals of hemoglobin. In the two photographs, the spots were in the same positions, but the intensities were significantly different. He then realized that he could obtain the three-dimensional structure of a molecule containing thousands of atoms. The multitude of three-dimensional structures available now from databases are the result of this unique moment of insight.

9

The Combs of the Wind: Unweaving the X-ray Rainbow*

I am purposely going to take a wide detour to clear your mind a little bit. The last few essays are conceptually the *inner sanctum* of crystallography and we need to take a break. The vast majority of educated people in the world know who Ferdinand Magellan (1480–1521) was. They all would associate his name with the explorer and navigator who went around the globe for the first time. However, some of these people will also have notions that the leader of the first sailing expedition around the world was killed in a skirmish with the native inhabitants of the island of Mactan in the Philippines. The 1974 edition of the *Encyclopedia Britannica* that I inherited from my father says that the voyage "was completed by others." After the perils and calamities that the crew of *Victoria* suffered after they left the Philippines, I do not think that bringing the vessel safely back to Spain is a feat that should remain anonymous. This is particularly so when one realizes that Portugal, a declared enemy of Spain, not only commanded all the trading seaports but also controlled the maritime routes of the time around India and Africa. I mention this because all Spaniards know that only seventeen people completed the expedition aboard the spice-laden *Victoria* (an 85 tons

---

* Title inspired by the book *Unweaving the Rainbow: Science, Delusion and the Appetite for Wonder* by R. M. Dawkins (b. 1941) (Houhgton Mifflin Co. 1998). The book title itself related to the poem *Lamia* (1820) by J. Keats where this expression appears for the first time.

vessel) and the person who brought the leaking ship safely to port was the Basque sailor Juan Sebastian de Elcano (1487–1526).[1] In the 1994 edition of the *Britannica* his achievement is properly recognized. As a recompense for this mighty feat, the King of Spain engraved in his coat of arms the legend "*Primus circumdedisti me*" (the first to go around me). This legend is proudly displayed in the coat of arms of his hometown, the Basque town of Getaria (*Guetaria* in Castillian spelling), in the region of Guipúzkoa in northern Spain. Incidentally, this tiny Basque village is also the birthplace of Doña Pepita Embil, mother of the world-famous tenor Plácido Domingo.

As a person born in the Castilian plateau, I have always been mesmerized by the sea and especially by the *Cantabrian* sea. Although not officially known by this name, it is the portion of the Atlantic that bathes Southwestern France, the Basque provinces of Spain, and further west Cantabria (Santander) and Asturias. Officially it is the Bay of Biscay (a name probably derived from the Bay of the Basque). Hidden in this rugged landscape where the mountains meet the sea there is a monument to Juan Sebastian de Elcano in his hometown of Getaria. The tiny village is barely visible behind a promontory that is known as "*El ratón de Getaria*" ("The mouse of Getaria") because it resembles the hump of a giant mouse. In this remote village of ancestral whale-hunters there is no place to go but to the sea. To this day, every four years on January 17 there is a symbolic procession of 18 solemn, candle-carrying men from the fishing port to the main parish of San Salvador to re-enact Elcano's feat.

Approximately thirty kilometers east, along the coast and near the capital of Guipúzkoa (Donostia in Basque—San Sebastián) is another of my favorite spots. Perched in different rocks there is a group of three sculptures in wrought iron by the Basque sculptor Juan Eduardo Chillida (1924–2002) entitled *The Combs of the Wind* ("*Los Peines del viento*," **Plate 9.1**). These iron sculptures combine, in a unique open-air environment, all of the elements that have made Chillida's sculpture world-recognized: a craftsman's feeling for the raw materials and the sharp contrast between the ruggedness of the immobile iron and the openness and incessant motion of the surroundings.

Believe or not, we will reprise the *"The Combs of the Wind"* at the end of the essay. Please, continue on.

Any measurement in science comes associated with a certain amount of error and the data obtained in crystallography are no exception. The introduction of a heavy tuba in a close-knit arrangement of musical instruments will probably disturb the original positions and you may have to try the diffraction experiment a few times. Our audience may be tired after several hours of intense listening and their reported values may not be properly correlated with the previous ones. In reality, protein crystallographers observed that in many cases proteins modified with heavy atoms do not give data of comparable quality to the native set and so, in the early days, as many as three-to-five heavy atom sets had to be obtained to calculate reasonable phases. Not to mention the fact that crystals mounted in capillary tubes decayed with time in front of the X-rays, making it difficult to measure accurately the subtle differences between the native and the substituted proteins.

What can we do to enhance the signal? Nature's subtlety and human ingenuity come into play again. Atoms, almost like people, are very sensitive to the presence of stimulating energies near their natural frequencies of excitation. When they sense such a frequency they behave in an "anomalous" way and their response to the incoming radiation is enhanced. Like musical instruments, atoms will resonate to the right frequencies and will send stronger signals. Can we measure and utilize this anomaly? Can we tune-in to optimize this signal? What atoms have the best anomalous response in the presence of the appropriate radiation? Ever since the first protein structures were determined, the community of protein crystallographers was thinking about how to exploit this property of atoms to enhance the signal of the diffraction experiments, and also to add additional information to the diffraction data.

Many macromolecular crystallographers (including myself) come with a background in the physical sciences. A well-known figure is F. H. C. Crick (b. 1916), co-discoverer of the DNA double helix with J. D. Watson (b. 1928). We were attracted to the field by the possibility of applying methods and ideas of the physical sciences to the biological sciences. This strong connection with the theories and technologies emanating from the hardcore physical sciences (particle accelerators, detector technology, X-ray optics, ...) helped protein crystallographers solve most of the problems indicated before. A new source of X-rays, different from the conventional sources using copper targets, began to appear in the early seventies—synchrotron rings. (See Part VI on new technologies for details). It is as if a modern Cyclops has given crystallographers a new torch to illuminate their samples that has all the desirable properties. It is extremely intense in a small area, tunable, and made up of internal pulses.

In addition, protein crystallographers have another valuable group of colleagues and techniques. These are the ones associated with the production of protein samples in quantities large enough to be able to pursue crystallographic studies (approximately 5–50 mg). These techniques evolved from methods to manipulate DNA fragments and introduce them in bacteria, allowing the bacteria to produce extraneous proteins for laboratory study (see Chapter 21, *"The Soft Engineers"*). In particular, crystallographers and their molecular biology colleagues were able to produce proteins with the atom sulfur replaced by selenium, a rare element in biological tissues. These developments combined have revolutionized protein crystallography. It has developed from an arcane specialty to a mass production method that promises to uncover thousands of new protein structures in the next few years (see Chapter 26, *"Genomics, Proteomics, and the Essence of Life"*). How is this possible? How does it work?

I'll try to guide you through the answers to these questions by slightly changing our magical concert hall blown by a strong and mysterious zephyr. The extremely intense X-rays from the new sources can be easily disguised as a very strong wind that will make our concerts very short because we can measure the tenuous sound of the instruments very rapidly. The replacement of sulfur—a relatively rare atom in proteins—by selenium would be analogous to replacing an instrument in the stage (i.e., a violin) by another violin of essentially the same size but containing a very special and conspicuous component, perhaps a golden bridge or a metallic peg. This replacement will produce minimal disturbances in the pre-existing arrangement of the musical instruments and will reduce errors for distorting the crystal lattice (lack of isomorphism).

But, you might argue, this is only one replacement and besides it may not be big enough. Don't we need a minimum of two? Of course we do, but the rest is taken care by the uniqueness of synchrotron rings. I will now bring back the sculpture that can "comb the winds" and pretend that if I put it before the stage, it can change the "color" of the wind in our concert hall. Suppose that we can tune-in the wind to the color of the golden bridge. Then the violin produces a special sound, which can be easily detected and measured by our audience. This could reduce all the tedious experiments involved in the determination of the phase by **MIR** to three rapid auditions, with the wind tuned to three different colors: right before we reach the bright color of the bridge, at the peak of the match, and well away from the transition. If the gold were to glow when the right color is reached, then you could collect the measurements from the audience right before the glow, at the peak of the glow and away from the glow. In this day of electronics, one can envision the audience having some push buttons to report immediately to a master computer their position and the sound measured, reducing the time of the experiments dramatically. The technique is called **M**ultiwavelength **A**nomalous **D**iffraction (**MAD**). You can read more about the impact of this method in a later essay (see Chapter 16, "*The Lunar Element and Our Collective MADness*"). The data from these three experiments combined are equivalent to the one obtained from the **MIR** experiments.

Now you probably realize that my "*grand detour*" at the beginning had a purpose. I wanted to introduce indirectly a device that would permit us to conduct a *gedanken* experiment in our magical concert hall. I could not think of anything better than a beautiful sculpture that combines together the rugged beauty of the mountains, the unique texture of forged iron, the vistas of the Cantabrian sea, and the misty and salty wind blowing through the Basque landscape. I hope you enjoyed the tour.

10

Prof. M. G. Replacement's Vision: Homage to Prof. M. G. Rossmann

Perutz and his co-workers had shown the power of the method of Multiwavelength Isomorphous Replacement (**MIR**) to obtain the phase angles associated with certain reflections in the diffraction pattern of horse hemoglobin in 1953. They were able to diffuse several compounds containing the heavy atom mercury (Hg) into the crystals of native, unperturbed, horse hemoglobin and measure data from the native and mercury-soaked crystals. These multiple data sets allowed them to obtain phases for special subsets (i.e., planes) of reflections in the native crystals. Within a few more years, the application of this method to the complete data of myoglobin (at a resolution of 6 Ångstroms in 1957.) and later hemoglobin (5.5 Å, 1959), unveiled the first three-dimensional structures of proteins solved using the **MIR** method. However, the application of this method to other proteins in later years became rather elusive. It was difficult to prepare derivatives of the protein crystals by adding other heavy metals, as had been shown in the globins, and this limitation appeared to be a major impasse. Consequently, by the end of 1960 only the two related oxygen-carrying proteins, myoglobin and hemoglobin had given away their internal structure. There was a sense of frustration in the small community of protein crystallographers.

According to Prof. M. G. Replacement,[1] it was at the International Congress of Crystallography held in Cambridge in 1960 when a way to bypass this problem occurred to him. Two threads of thought lead him to the moment of insight. First, it was beginning to be documented that the same protein molecule could be crystallized in different geometrical arrangements. The same molecular object was sometimes arranged in various ways (i.e., packed in different crystal forms with different unit cells). Second, large proteins were often made up by clustering identical units in discrete symmetrical arrangements such as dimers, trimers, tetramers, hexamers, and even larger aggregates. Moreover, these clusters of identical pieces would frequently occupy the asymmetric unit of the crystal providing additional symmetry, or non-crystallographic symmetry. It occurred to Prof. M.G. Replacement that perhaps these two properties could form the basis for a computational method for solving protein structures, without the need to use the unpredictable alchemy of heavy atoms. How will this work? In which way is this approach experimentally simpler than the use of several heavy atoms (**MIR**) or anomalous scatterers (**MAD**) discussed previously?

For starters, the method uses only one single data set, obtained from the native or original crystals. Consequently, it will reduce the time required to collect the data and also the errors associated with the processing and comparison between the different data sets. In its simplest conceptual description, the method of **M**olecular **R**eplacement (**MR**) seeks to orient and locate the model of a similar protein, obtained from other crystals, in the unknown crystal. In this context, the method is a direct descendent of the crystallographic methods referred to as "Patterson" methods, pioneered by the American crystallographer A. Lindo Patterson (1902–1966) and David Harker (1906–1991). Think of the known protein structure as an asymmetrical three-dimensional object (i.e., a right shoe). It is assumed that the unknown crystal contains a different but closely related protein (another similar right shoe for the same person), of which the orientation and exact position are unknown. You can see that the problem can be separated into two stages:

i) Find the orientation of the model in the unknown crystal (three angular parameters).

ii) Keeping the orientation constant, move the model back and forth until you find the approximate position in the crystal (three position parameters x,y,z).

Once you have values for those six parameters (three for orientation and three for position) you can calculate approximate values for the phases of the reflections in the unknown protein, and careful refinement will do the rest. Because you already knew the approximate structure of the unknown protein, you have conceptually and mathematically reduced the number of variables from several thousands to six. Not bad, right?

If there was no pre-existing model of the unknown protein, Prof. M. G. Replacement argued that you could use only *one* isomorphous heavy-atom derivative to obtain crude electron density maps. Then you could average the initial values among the repeated copies of the protein in the unique portion of the crystal (i.e., its asymmetric unit) to improve your signal, thus refining your phases. He envisioned that these methods would be particularly useful in the solution of the structure of viruses, which had been shown to me composed of many identical protein subunits (see Chapter 22).

These conceptually simple ideas required a mathematical foundation and a series of computer algorithms and programs that could make it feasible to calculate initial models and phases for the unknown protein structures. With the computers of the time this was not a small task and Prof. M. G. Replacement has confessed to "frequent nocturnal orgies with various computers"[2] to develop and test the initial programs. It has taken approximately twenty-five years for the method of **M**olecular **R**eplacement (**MR**) to become a routine in macromolecular crystallography. As the number of known protein structures (or folds) continues to increase, its future impact will be staggering.

Routine application of **MR**-methods in protein crystallography has become so widespread that many older practitioners in the field, or even some of the novices, may experience a sense of *déjà vu* when reading the methodology section of many crystallographic papers. They have repeatedly encountered sentences like: "The structure was solved by the method of MR as implemented in the program suite [...]." In a closely related context they see "The initial phases were improved by non-crystallographic electron density averaging between the N copies in the asymmetric unit [...]." Yet the ideas, the concepts, and even the terminology were *terra incognita* only thirty years ago. I do not intend to explain the method here in any more detail, nor do I intend to show statistically how many macromolecular structures have been solved by the explicit or implicit use of those ideas. Those facts are available from textbooks and Internet sites.[3,4] I just want to pay homage to the person who, approximately thirty years ago, had the intellectual vision of using the conservation of the three-dimensional folds of proteins, and the "redundancy" of information contained within the asymmetric unit to aid in the structure solution of macromolecular structures.

Prof. Molecular G. Replacement was born in Frankfurt, Germany on July 30, 1930, and immigrated to England with a member of his family when he was nine years old. He obtained his B.Sc. from the University of London in 1950, his M.S. in 1953, and a Ph.D. in Chemistry from the University of Glasgow in 1956 under the mentorship of J. M. Robertson. He did postdoctoral research from 1956–1958 with Prof. W. Lipscomb in Minneapolis. Inspired by a lecture by Dorothy Hodgkin to work on the determination of the crystal structure of biological macromolecules, he returned to England to work with Max Perutz on the structure of hemoglobin at the Medical Research Council (MRC) during the exciting years spanning from 1958 to 1964. From then on, his ideas, programs, structures and papers have had a tremendous influence in the field of macromolecular crystallography.

He married a remarkable woman, Audrey Pearson, in 1954 and they have three children (Martin, Alice, and Heather). Many old time crystallographers will remember the very first "cartoon" sketches of lactate dehydrogenase (LDH) (**Figure 10.1**). Following a suggestion from Anders Liljas (a Swedish post-doc crystallographer in the lab), they were drawn by Audrey to simplify the wanderings of the polypeptide chain in space. Jane Richardson later popularized more

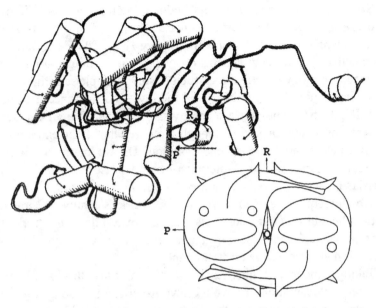

Figure 10.1. Original cartoon drawing of LDH. Schematic representation of the meandering path of the polypeptide chain in lactate dehydrogenase as originally conceived and drawn by Audrey Rossmann. By convention, the arrows represent β strands and the cylinders α helices (see Chapter 20 for the atomic structure of these features). Inset represents the arrangement of the four LDH molecules (monomers; each fish represents one molecule of LDH) in lactate dehydrogenase, forming a tetramer with three perpendicular, intersecting, two-fold axes of symmetry (222 symmetry). These are labeled **P** (in the West direction), **Q** (towards the viewer) and **R** (vertical). (Inset drawn by the author and published in *Journal of Molecular Biology*;[5] reproduced with permission, courtesy of Academic Press.)

artistic versions of those sketches and for a while they were essential images of any new protein fold. Those drawings that today are so commonplace in all kinds of colors, shades and hues have their origin in those crude hand-drawn diagrams. Incidentally, "Aud" is also a superb potter and I treasure my coffee mug, our dinnerware and any other piece that comes from her creative mind and her amazing pair of strong hands.

Some may wonder why this sudden interest in recognizing Prof. M. G. Replacement. I must confess that his 65th birthday is only partly the reason (**Plate 10.1a,b** and **Figure 10.2**). The idea originated when I recently saw a paper in one of the leading journals of molecular biology describing a structure solved by phase refinement and electron density averaging, and there was no reference to Prof. M.G.R. At first I thought it was pretty sad, and that is why I decided to write this essay. On second thought I realized that in the end it may the greatest honor to pass into oblivion. Do we quote Sir Isaac Newton or even Galileo Galilei every time we use the principle of inertia? Do we refer to the author of the *Principia* every time we use any of his equations? Certainly not. His work is totally integrated within the fabric of our science and our culture. Similarly, the work of Prof. M. G. Replacement is now, and forever more, part of the framework of macromolecular crystallography.

Taking some literary liberties, I would like to finish this brief homage with an adaptation of a well-known American folk song. I leave it to the reader to find out the first and last names of Prof. M. G. Replacement, his favorite hobby (I already disclosed the name of his wife and partner), and the name of the river that crosses the city in Indiana where both have lived since 1964.

M... sail the boat ashore, Hallelujah
M... sail the boat ashore, Hallelujah

Audrey help to trim the sails, Hallelujah
Audrey help to trim the sails, Hallelujah

Wabash river is chilly and cold, Hallelujah
Freezes the body but not the soul, Hallelujah

Wabash river is deep and wide, Hallelujah
Milk and honey on the other side, Hallelujah

1. Ming Luo	56. Yasuo Hata	112. Adrian Mulholland
2. Jun Tsao	57. Keiichi Fukuyama	113. Chris Colbert
3. Fei Gu	58. Xiaoxia Zhao	114. Wladek Minor
4. Lorna Ehrlich	59. Kumiko Tsukihara	115. Sanjeev Munshi
4a. Carol Carter	60. Tom Smith	116. Abelardo Silva
5. Cory Momany	61. Dino Moras	117. Dick Verduin
6. Rui Zhao	62. Roger Burnett	118. Mavis Agbandje-McKenna
7. Uli Grau	63. Andy Muckelbauer	119. Ulrike Boege
8. Veda Chandrasekar	64. Prasanna Kolatkar	120. Andy Prongay
9. James Owen	65. Liwen Niu	121. Christopher Woenckhaus
10. Cele Abad-Zapatero	66. Seungil Han	122. Doug Scraba
11. Katey Owen	67. Nino Campobasso	123. Gail Ferstandig Arnold
12. Steve Harrison	68. Diana Tomchick	124. Sandra Gabelli
13. Roland Rueckert	69. Abhinav Kumar	125. Colin Parrish
14. Robert Huber	70. Noel Jones	126. Ed Arnold
15. Dietrich Suck	71. Di Xia	127. Jean Hamilton
16. David Davies	72. Joe Ferrara	128. Keith Perry
17. Jack Johnson	73. Rob McKenna	129. Nino Incardona
18. Michael Rossmann	74. Bill Cramer	130. Barney Axelrod
19. Peixuan Guo	75. Mike Becker	131. Tom Hess
20. Andrea Giacometti	76. Richard Schevitz	132. Steve Tracy
21. Sharon Wilder	77. Mark Hermodson	133. Brian Bowman
22. Sukyeong Lee	78. Antonio Llamas-Saiz	134. R. Chandrasekaran
23. Beverly Heinz	79. Carol Post	135. Joe Krahn
24. Bill Farrar	80. Marvin Hackert	136. Bill Ray
25. Rik Wierenga	81. Mitchell Guss	137. Peter Willingmann
26. Alex McPherson	82. David Belnap	138. John Burgner
27. Graheme Williams	83. Arlett Harris	139. B. V. V. Prasad
28. Cynthia Stauffacher	84. Zhongning Yang	140. Hans-Jürgen Hecht
29. Paul Kaesberg	85. Wilfried Schildkamp	141. John Tesmer
30. Jeff Bolin	86. Kenton Longenecker	142. Tom Hurley
31. Carla Little	87. Peiguang Xie	143. Hank Weiner
32. Steve Warren	88. Michiko Konno	144. Takashi Yamane
33. Barbara Harris	89. Akella Radha	145. Klaus Herrmann
34. Jodi Muckelbauer	90. Nobuo Tanaka	146. Laci Kovari
35. Christine Muchmore	91. Paula Fitzgerald	147. Cyndy North
36. John Erickson	92. Bill Stroud	148. Toshio Akimoto
37. Jin-Huan Liu	93. Tim Baker	149. David Blow
38. Sherin Abdel-Meguid	94. Michael Chapman	150. Peter Williams
39. Tomitake Tsukihara	95. Sergei Strelkov	151. Adam Zabell
40. Jin-bi Dai	96. Liang Tong	152. Steve Walker
41. Maitreyee Chandra	97. Alexander Tong	153. Mittneen Williams
42. Zhiwei Che	98. Yizhi Tao	154. Anders Liljas
43. Zhu Zhu	99. Hao Wu	155. Paul Chipman
44. Holland Cheng	100. Irwin Tessman	
45. Marcos Oliveira	101. Jie Yang	
46. Guy Diana	102. Jia-huai Wang	
47. Shamsa Faruki	103. Barbara Sherry	
48. Jordi Bella	104. Giuseppe Terracina	
49. Terje Dokland	105. Andy Fisher	
50. Sangita Sinha	106. Guoguang Lu	
51. Tim Schmidt	107. Bonnie McKinney	
52. Jeff Greve	108. Jim Ely	
53. Young Jo Farrar	109. Andrea Hadfield	
54. Ed Lattman	110. Steve Muchmore	
55. Mark McKinlay	111. Richard Kuhn	

Figure 10.2. A name list to the identity of friends and colleagues of Prof. M. G. Rossmann on the occasion of the symposium entitled *New Directions in Protein Structure-Function Relationships* organized to commemorate his 65th birthday (see Plate 10.1).

FRODO,
The Electronic Hobbit:
Molecular Graphics

Dedicated to Alwyn Jones

From early childhood, John Ronald Reuel Tolkien (J.R.R. Tolkien, 1892–1973) was fascinated with languages. When he was five, his mother—who was fluent in Latin, French, and German—taught him to read in all three languages plus her native English. Fatherless since 1896, the family lived in a small rented cottage in the hamlet of Sarehole by the Cole River, far from the smokestacks and soot of Birmingham. The quiet meadows and streams of Sarehole were a haven for Ronald and his younger brother Hilary. There his mother introduced them to botany and inspired in them a love for plants, trees and the beauty of natural landscapes. Nonetheless, change again came to his life abruptly. His mother died in 1904 and the brothers were left under the guardianship of a Catholic priest, Father Francis Morgan, who had a tremendous influence on his education and his life. Tolkien graduated from King Edward VI school in Birmingham and won an award to attend Oxford University. His interest in and passion for languages led him to study philology, specializing in the literary and linguistic tradition of the English West Midlands. This study requires extensive knowledge of Anglo-Saxon (or Old English, as in Beowulf), Middle English (the language of Chaucer), and Finnish, Icelandic, Norse and Germanic mythologies and folklore. He was Professor of Anglo-Saxon at Oxford and a Fellow of Pembroke College from 1925 to 1945. Later he became Professor of English Language and Literature and a Fellow of Merton College from 1945 until his retirement in 1959.

Tolkien's academic achievements are inextricably linked with the creation of two multi-faceted, highly imaginative epic stories which had a tremendous influence on the youth of the 1960s all over the world, and whose effect still reverberates today. In 1937 he published *The Hobbit*,[1] which received high acclaim as a fascinating children's story. The main characters were a "hobbit" named Bilbo Baggins and a wizard of sorts named Gandalf. Tolkien later wrote that the origin of the word hobbit seems to be: "a worn-down form of a word preserved more fully in the language of Rohan: holbyta or 'hole.'"[2] What is a hobbit? Here is a brief description, paraphrasing Tolkien's inimitable prose.[3]

Hobbits are beardless little people, approximately one half of our normal height. Their only magic power is their ability to disappear quietly and quickly when large people like us come stumbling around and making heavy noises like elephants, which hobbits can hear from a distance. They are a bit chubby in their stomachs and like to dress in bright colors (mainly green and yellow). Although they never wear shoes, they never walk—in a way of speaking—"barefooted" because their feet grow natural leathery soles and thick and fluffy hair, similar to the curly stuff on their heads. Hobbits have long, nimble, brown fingers; generous, friendly looking faces, and laugh profound "fruity" laughs. This is particularly true after dinner. This should be enough to continue on.

The illusion of hobbits as calm, simple, people capable of heroic feats caught on quickly and Tolkien was asked to write more adventures of Bilbo Baggins. *The Hobbit* had ended with Bilbo keeping a ring that he had found during his fight with Gollum, and living happily in the Shire: the idyllic part of Middle-earth where the hobbits lived and that scholars have related to the Sharehole of Tolkien's childhood.[4] The author had no desire to write a sequel. Instead, *The Fellowship of the Ring*, the first volume of the epic trilogy *The Lord of the Rings* was published in 1954. Soon after, the next two volumes appeared, *The Two Towers* and T*he Return of the King*. The completed work was a mythological world of monumental proportions in which Tolkien had given life to creatures, kingdoms, wars, calendars, climates, places, landscapes, and seasons to give flesh and blood to the languages spoken by the people of Middle-earth: humans, elves, trolls, goblins, giants, dragons, ents, balrogs, orcs. The hero was Frodo, heir and nephew of Bilbo Baggins, who together with his friend Sam and other companions of the fellowship undertake a quest to destroy Sauron, the master of the evil ring that Frodo had inherited from his uncle. The appeal of an innocent, gentle creature succeeding in destroying the forces of evil against all odds, in an unspoiled landscape of pristine forests, mountains and lakes was enormous. By 1967 *The Lord of the Rings* had been translated into nine languages with an estimated readership of fifty million people. The graffito "FRODO lives!"[5] appeared in the New York subway as testimony to a cultural phenomenon that had opened a magic wonderland of places, characters and events unhindered by the prosaic incidents of our every day lives. Tolkien had transcended the arcana of scholarly research in obscure languages to create a universal allegory of the constant struggle of good against evil, with strong environmental overtones. A recent motion picture has brought all these places, characters and imaginary events to larger audiences and younger generations in spectacular settings and imagery.

There is, however, another, less known, incarnation of the most famous of Tolkien's characters: FRODO, the electronic hobbit, who had its origins in 1976. Its appearance on the scene was preceded by other hobbits such as GRIP, and grew concurrently with FIT and other lesser-known hobbits. Whether the younger generations believe or not, at that time most protein models were built starting from a C_α tracing obtained from markings on tracings of electron density maps drawn on small sheets of transparent plastic, sometimes stacked up as "mini-maps".[6] From an initial set of guide coordinates, detailed atomic models were built at a large scale in a Richards optical comparator (known in the trade as "Richards Box" or "Fred's Folly") using Kendrew model parts (**Figure 11.1**).[7] Glass or plastic windows had to be drawn by hand with tracings of the electron density contours at the appropriate scale (2 cm = 1 Ångstrom). The scale was important; the scale originally planned for the myoglobin structure was revised downwards when it was realized that the floor of the modeling room would not bear the weight of the finished model. Atomic coordinates were laboriously extracted from these wire models by tedious and often inaccurate protocols.[8] Although maps could be hand contoured and placed in the Richards Box for analysis within a reasonable period of time (days to weeks), there was an urgent need for a computerized method that would allow the fitting of an atomic model to the experimental electron density map, and which would remove the tedium and inaccuracies from macromolecular model building and refinement.

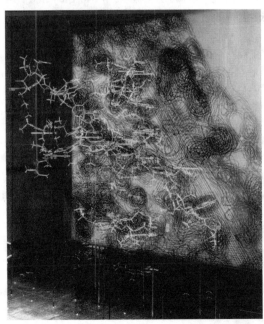

Figure 11.1. A view of the wire model of the protein coat of Southern Beam Mosaic Virus (SBMV) built at Purdue University in the late 1970s and early 1980s. In the foreground one can see the metal parts making up the atomic model, supported by metal rods screwed onto a plastic surface below (not seen) containing a coordinate grid (x and y; z is the height from the surface). In the background one can see the electron density map hand-drawn on large glass windows.

The idea of using interactive three-dimensional computer graphics for this purpose was floating around the community. This problem was at the absolute cutting edge of what was possible with the hardware at the time. Several projects were underway to achieve this goal, including those of Ed Meyer Jr.; Fred Brooks Jr.; Dave Berry; and Bob Diamond. I leave it to the historians to investigate the technical and human details.[9,10] Somehow, Drs. J. Gassman and R. Huber found a bright young Welsh computer scientist who was interested in living in Munich to develop such a tool. The reasons behind the interest to reside in Germany by this scientist have been explained.[10] Circumstances aside, Gassman and Huber encouraged him to make a program useful for the routine operation in a protein crystallography laboratory. Tradition has it that the original program sent data back and forth between a PDP-11 and a Siemens 4004, in a computing environment where many of the programs were named after different hobbits. It was only natural that the central program would be named after the most famous of all the hobbits in Tolkien's trilogy. For obvious reasons, the test version used most of the computing cycles and was initially called SAURON.

As a computer graphics program, FRODO made its appearance in the protein crystallography community more than twenty years ago in 1978.[11] As for myself, I got to know FRODO very well in 1981 during three beautiful weeks of immersion within the incomparable Swedish spring. Unfortunately, Vithorita, pregnant with our first child, could not join me. My friendship with FRODO developed during many nocturnal model-building sessions at the old Wallenberg Laboratory next to the ancient city castle in Uppsala, Sweden. I must confess that we had our crises, but he was certainly a very friendly hobbit. I was the one to blame for every crisis. Quite often, I failed to understand his prompts or suggestions, and many times his cues made no sense to me. He was always patient, effective and obedient.

You could **CHAT**[*] with him via a keyboard but the most effective way to communicate was with a tablet and a pen that would allow

---

[*] Actual FRODO commands are in bold capital letters.

you to pick and identify atoms, and select different commands from a **MENU** on the screen. Obedient to the **GO** command, FRODO would display for you a certain volume of electron density. Using well-designed commands, you could tell him to **BREAK** certain bonds and cut the protein chain into pieces. These pieces could then be moved with six degrees of freedom (**FBRT**) to make them fit into the three-dimensional electron density maps, which could be rotated at will with dials or joysticks (**Figure 11.2**).

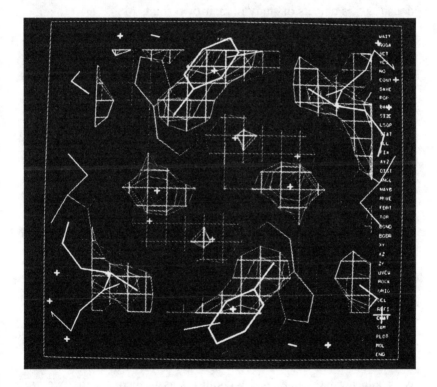

Figure 11.2. A real screen of the original FRODO program. Although lacking colors, FRODO offered to the user many interactive conveniences on a "menu" displayed on the right hand side of the screen. Many of these options are the commands discussed in the text (presented in bold).

FRODO did not know any protein chemistry, or if he did, he would not explicitly tell you so. It was you who would organize those constellations of points in space into a meaningful protein chain by using the **REFI**nement command. He would faithfully apply the rules of chemistry to certain **ZONE**s of your spatial points that were covered by your electron density contours. This was a tremendous help when trying to fit those old electron density maps. FRODO was also very handy at modeling exercises by allowing you to create **MOL**ecular objects that you could use either as background while fitting electron density or as objects of study in their own right.

For some time, the rumor (joke) floated in the community that the only documentation for FRODO was *The Lord of the Rings*. This might have been true, but in his own humble way FRODO proved to be a very useful hobbit and was the ancestor of many other electronic hobbits that are now well settled in our computer underworld. In addition, his faithful friend SAM was always available to insert or delete residues, create a sequence, and do all the necessary bookkeeping so that in the end everything was **SAVE**d in the disk with "amazing speed" and accuracy. During my visit, FRODO lived in a simple VAX 11/750 computer and his commands drove a Vector General VG3400 graphics display. Later he lived inside many other boxes or hobbit-holes in many other countries. His performance improved as his electronic eyes and hands improved, permitting us to view unimaginable shapes and forms and to examine atomic continents, islands, and landscapes of indescribable complexity and beauty. Following his original insights, we can now see atomic crevasses and caves, canyons, rivers, mountain ridges, and valleys in different and vivid colors, subtle hues and delicate shades. FRODO opened for us an atomic underworld that was previously beyond our reach. He introduced us to an atomic Middle World that we could not have imagined without his assistance and that we are just beginning to explore, appreciate and understand using more modern tools (see **Plate 11.1**). The image on **Plate 11.1** also dramatically illustrates the final result of all the crystallographic calculations using the Fourier sums: an electron density map. The contouring lines (white) that surround and engulf the chemical backbone of the polypeptide chain is the electron density map. Note the fit of the chemical structures to the shape of the electron density: blue represents nitrogen atoms, red is used for oxygen and yellow for sulfur; carbon is white. Hydrogen atoms are normally not drawn in protein structures. The central part of the polypeptide chain has been depicted in solid tubes; background amino acid residues are drawn with simpler lines. On the right one can see some of the multiple options available on the "menus" of today's computer graphics programs. To the lower left a discrete electron density peak corresponding to a water molecule has been encircled. One amino acid consists of a nitrogen, a central carbon (C_α), and a C=O group (for example, Alanine217). As hydrogen atoms cannot be seen at this resolution, they have not been included for clarity. Different amino acids have different groups (side chains) emanating from the central carbon, except the amino acid glycine that does not have any extension. The peptide bond linking two amino acids together can be represented as –(C=O)–(NH)– (…white=red-blue…). See **Glossary** for the exact chemical diagrams. Clockwise from the central, sulfur-containing (yellow), residue (Cysteine215) you can see six different amino acids forming approximately a circle in the center: Cysteine215-Serine216-Alanine217-Glycine218-Isoleucine219-Glycine220-Arginine221.

One could argue that there are no malicious villains in our atomic Middle World. There are no Dark Riders or Ringwraiths trying to prevent FRODO from destroying the evil ring. Yet, we routinely encounter and study molecules with pathogenic or curative properties in our crystals, and a major part of our time is spent trying to understand their interactions with themselves and with other molecules. With our efforts we are trying to defeat the evil forces of disease, pain and deformity and our operational domain is the atomic Middle World that FRODO unveiled for us. There are parts of these atomic creatures that we cannot see or cannot fit well in our electron density maps, and that chase us in our sleep like the Dark Riders chased after FRODO and his friends. However, our true Gollum, Shelob, and Sauron are uncertainty, lack of knowledge, and most especially bias and disorder. Those restrictive forces will always be with us. In the meantime, FRODO will live on in the heart of those of us who once upon a time built protein models using mechanical parts and read the coordinates of our structures using a two-dimensional grid and a plumb line. He did so many things for us, he was such a good friend...

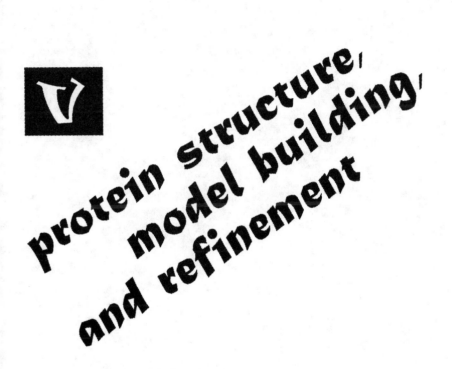

V

protein structure,
model building,
and refinement

12

Only Refined Proteins Go to Heaven

The fruit of the labor of crystallographers in general and, more specifically, protein crystallographers is a publication in a professional journal, with the work reviewed (or refereed) by colleagues knowledgeable in the field. The outline of these structural papers is very similar: Introduction; Materials and Methods (where the preparation of the sample is discussed as well as the crystallographic techniques employed); Results, presenting the structure, and finally another section typically discussing the implications of the work for the specific chemical or biological problem. I recently was reviewing a typical manuscript in the field and reading the standard sections: Introduction, Materials and Methods, Results and Discussion, Structure Solution and Refinement, with some discussion as to the biological implications of the refined structure. Something caught my attention after the refinement section:

> "The coordinates have been deposited in the Protein Data Bank with *ascension* code…".

I do not recall the exact entry code. However, "*ascension*" was obviously a misprint. The authors did not mean to write down *that* word. Yet, I could not resist the pun. Was this a Freudian slip? Had the authors implied that after so many years of hard work on that particular protein, and after the structure had been solved and refined, their results had to ascend to some special "heaven of refined proteins?" Well, yes and no.

The Protein Data Bank (PDB) had its origins at a meeting of macromolecular crystallographers at Cold Spring Harbor (Long Island, New York) in 1971. At this meeting, the need for a center to archive the structural data from macromolecules was formally discussed. Dr. Helen Berman, Dr. Walter Hamilton, and Dr. Joel Sussman, among others, began to take an active role in planning the logistics for the incipient center. The PDB (as it was later informally baptized) was established at Brookhaven National Laboratory (Upton, Long Island, NY) in 1971 as a depository for the coordinates of the protein structures to be determined by macromolecular crystallographers.[1] As of this writing, the PDB has been around for over three decades during which it has been a valuable resource to an international community of researchers, educators, students and professionals in a wide variety of scientific disciplines, from biochemists and health professionals to materials science and environmental investigators.

In parallel with the development of the PDB and growing at a much more rapid pace, there is a myriad of sequence databases containing the nucleotide and amino acid sequences included in the genome of a multitude of organisms. In the past, the elucidation of the amino acid sequence of just a single protein by chemical means was a daunting task. The decoding of the exact amino-acid sequence of bovine insulin (two chains, A and B, of 21 and 30 amino acid respectively) between 1949 and 1955 by Frederick Sanger (b. 1918) was one of the key discoveries in the dawn of molecular biology and the unraveling of the genetic code. Development of other ingenious chemical methods also by Sanger, Paul Berg (b. 1926), and Walter Gilbert (b. 1932) allowed them to obtain the precise order (sequence) of nucleotides in the genetic material (DNA or RNA) of entire viruses (bacterial virus R17, 5000 units) or cell organelles (mitochondria, 17,000 units). Further improvements of these methods now permit automatic sequencing of larger fragments of genetic material, resulting in the sequencing of entire genomes at vertiginous pace. All of this sequence information is stored, maintained and annotated by modern "curators" and specialists in bioinformatics, often working in collaboration with multinational genomic projects. Of special interests is the Swiss-Prot databank in Geneva, Switzerland, now a part of the Swiss Institute of Bioinformatics (SIB, or Institute Suisse de Bioinformatique or ISB).

The serendipitous path that led to the development of Swiss-Prot, from its humble beginnings in 1986 to its current prominent position, has been engagingly described by its founder Amos Bairoch.[2] Although it began as a simple depository of information, the advent of the World Wide Web and of dedicated computer servers, combined with the ingenuity of gifted bioinformaticians, has transformed Swiss-Prot and all other databases into unparalleled resources for the analysis and dissemination of biological information. A notable example is the EXpert Protein Analysis SYstem (www.expasy.ch) available at SIB since 1993. The PDB specializes in structural (three-dimensional) information. However, the Internet and technological and personal links with the faster-growing sequence databases have transformed it into a critical reference for the understanding of biochemical and biological function.

In the mid seventies, the PDB was the depository of the few protein structures that had been solved (a few had even been partially or fully refined!) using the methodology used by the protein crystallographers of the time: single crystal X-ray diffraction combined with the soaking of heavy atoms into the crystals. At that time, indeed, only a few proteins made it up there. They were very few for several reasons, the most important of which was that it was still relatively difficult still to solve protein structures by the **MIR** or **MR** methods (see Chapters 8–10). The technology was still a mixture of craftsmanship and science. In addition, some crystallographers were reluctant to make their hard-won structural results available to the community at large for a variety of pretexts.

The situation began to change when the National Institutes of Health, which funds more of the research in this field than any other source, made deposition of structures in the PDB a mandatory requirement for renewal of research grants, and the journals specializing in crystallography (i.e., *Acta Crystallographica*), and the ones presenting macromolecular structures (*Science, Nature, Biochemistry, Journal of Molecular Biology*, and others) began to require that coordinates be deposited at the time of publication. From then on, the PDB has been flooded with new macromolecular structures that have to be "annotated", "processed" and eventually made public to the community of structural biologists gratis over the Internet or for a modest media fee.

In recent years, the problem has been compounded by an increasing number of macromolecular crystallographers using better instruments to collect superb data rapidly, and improved methods for structure solution and refinement. Thanks to the efforts and ingenuity of the worldwide community of crystallographers structural results from crystals of macromolecules are becoming easier to obtain, to document and to archive.

The PDB is not such an exclusive place anymore, it was never meant to be. It has become very accessible. In the last few years, in addition to increasing numbers of highly refined structures determined by X-ray diffraction methods, excellent coordinates of model structures have also been deposited in the PDB and even the ensembles of structures determined by NMR also ascend to the PDB. With the advent of the technology from the world wide web, all these structural results can now be retrieved within seconds, and displayed and analyzed on any computer workstation. I was astounded, when one of my molecular modeler colleagues displayed, within seconds, a structure that amounted to several people-years of effort. Similar retrieval operations are preformed daily by the hundreds of thousands all over the world.

Currently, the PDB is operated by a research collaborative arrangement among three U.S. institutions: University of California San Diego Supercomputer Center, Rutgers University in New Jersey, and National Institute of Standards and Technology, Maryland, under the directorship of Prof. Helen Berman, one of its early founders. There are mirror sites of this facility around the world. As of May 2002, there were well over 18,000 macromolecular structures archived by the PDB. The dramatic increase in the number of macromolecular structures deposited in the PDB, as well as the rise of the structural complexity of the individual entries, has been beautifully illustrated by David S. Goodsell in a recent cover of the *IUCr Newsletter* (**Plate 12.1**). The number will increase even more dramatically in years to come with the future structural studies on the myriad of macromolecules of commercial and biomedical interest unveiled by the work on the complete genome of the first flowering plant (thale cress, *Arabidopsis thaliana*),[3] insects (fruit fly, *Drosophila melanogaster*),[4] fermenting yeast (*Saccharomyces cerevisiae*),[5] animals (nematode worm, *Caenorabditis elegans*),[6] pathogens (*Vibrio cholerae*,[7] *Haemophilus influenza*),[8] and the first draft of the complete human genome (*Homo sapiens*).[9,10] There are now some 800 complete genomes in the public sequence databases.

I should emphasize the enormous gap between the number of proteins and genes for which the amino acid sequence is known (several hundreds of thousands available from databases), and the much smaller number of individual three-dimensional structures (several to about ten thousand) available in the PDB. This gap is the driving force behind a massive effort in the macromolecular crystallography community to accelerate the structure determination of proteins as much as possible. Such a concerted effort of research has the goal of obtaining the majority of the protein structures coded by the genomes of several key organisms in a few years and is referred to as "structural genomics."[11]

If all these structural jewels have not ascended to heaven, many of them have certainly been elevated to the realm of eternal "Ideals" or abstractions. For example, we can now have an image of "hemoglobin"—the oxygen carrying protein in the blood—with its helical contortions holding oxygen very delicately, like a feather, ready to give it away, and to pick up carbon dioxide many times per second. Like Plato's images of the mind, they are ideal representations of the atomic world beyond our reach. Their shadows appear full of color in our electronic Plato's Caves by virtue of a few dials and a cathode tube (**Plate 12.2**).

However, protein crystallographers, like other common mortals, have to deal not with the ideal folds but with the individual variants. These often are hard to crystallize, their crystals do not necessarily diffract well, and they take time to solve and refine. The crystallographer's joy and satisfaction must lie in collecting as many of these beautiful pebbles as possible, and in understanding the significance of every detail and variation of these atomic masterpieces in the context of their biochemical function and of the biological processes that they support[12] (**Plate 12.3**).

13

On the Size, Shape, and Texture of Globular Protein Molecules

n the original papers of Max Perutz describing the three-dimensional structure of hemoglobin, he used the term "visceral" to refer to the contorted, twisted and irregular form of this protein, when compared with the simple and orderly DNA double helix unveiled ten years earlier.[1] Proteins are made-up of small (less than twenty non-hydrogen atoms: C, O, N, S) chemical entities, called amino acids, connected together like beads on a string by a type of chemical bond called a "peptide" bond (see **Glossary**). This long chain folds in three dimensions in rather irregular fashion. Globular, as opposed to fibrous, proteins are basically a ball of yarn, wrapping around a pocket or "active site," which is were the chemical activity of proteins takes place. In the case of hemoglobin, the oxygen-carrying protein of most vertebrates, the poly-amino acid chain folds around an atomic basket that holds oxygen delicately, so that it can give it away when the conditions are right.

Many of the papers that followed, describing the models of the newly solved protein structures by crystallography, attempted to describe their overall size and shape by giving the dimensions of the parallelepiped (i.e., boxes) or ellipsoid (distorted spheres) enclosing them. As more structures became known, attempts were also made to fit these unique, miniature, three-dimensional objects to certain canonical forms such as paraboloids, and prolate and oblate ellipsoids of different characteristics. These terms refer to mathematical solids specifying different types of distorted spheres. These various terms reflect the difficulty of describing very irregular objects in the framework of geometrical concepts and dimensions that we use to characterize many of our day-to-day objects.

I have frequently been fascinated by the problem of mathematically describing irregular objects such as protein structures. How do you define sizes and shapes of highly non-isometric (i.e., having different sizes in the three main directions of space) objects? Perutz was correct. Like many organic forms in nature, protein structures are contorted and visceral-looking, quite unlike the symmetric geometrical objects manufactured by our work-a-day human technology. The next few paragraphs will be somewhat technical, but I promise you that if you bear with me, we will discover a very interesting secluded byway, not overrun with tourists. If you do not have the courage, you can skip to where the text is less dense.

Around 1987, I convinced C. T. Lin, a statistician colleague at Abbott Laboratories, to help me tackle the problem from a different viewpoint. We worked together for a couple of years and in 1990 we published a paper entitled: *Statistical Descriptors for the Size and Shape of Globular Proteins.*[2] I admit that the title is cryptic. However, the insight is simple. One can define two purely statistical parameters (descriptors) derived from the set of points that define a protein structure, which can be used to describe the size and shape of such an irregular object. Our analysis suggested that, in addition to size and shape, another parameter related to the texture or roughness of the protein surface was needed in order to fully describe these complex shapes.

We restricted ourselves to the most "globular proteins"—the ones with the roundest outlines. We computed the frequency distribution of all possible distances between C_α carbons within the protein structure and studied the resulting distribution. This means that we computed the distances between all possible atom pairs, and then we clustered them into different bins or intervals. For example, how many distances are between 6.5 and 7.0 Ångstroms apart? Using all atoms in the structure did not modify the conclusions of the work. As expected, these distributions were skewed, quite unlike the symmetrical bell-shaped Gaussian curve expected for random variables. There is a sharp spike at 3.8 Å corresponding to the well known $C_\alpha - C_\alpha$ distance between adjacent residues in the polypeptide chain, which we can ignore for the rest of the discussion.

It was immediately clear that either the median[*] or mode[**] of the distribution was an excellent descriptor of the protein (object) size. Both descriptors correlated very well with the molecular weight of the proteins in the sample or to their radius of gyration (R_g).

Using a statistical trick named the Box-Cox transformation,[***] one can transform the observed distribution to be like a true Gaussian distribution. The resulting parameter λ (with values ranging from 0 to 1) can be used to measure how far the original distribution deviated from an ideal Gaussian distribution, whose λ value is equal to 1. After some analysis we realized that the parameter λ was also a good descriptor of the shape of the original object, similar in character to the axial ratio (a/b) of an ellipse or ellipsoid.

It is well known that the ratio of surface area to volume is a reasonable (albeit size dependent) measure of shape. We defined a size-corrected shape descriptor by the expression:

$$s = (Surface\ Area)(R_g) / (Volume),$$

resulting in a dimensionless quantity s. In physics, a dimensionless quantity is one that does not have any associated units. In this case, s

---

[*]   The median is the value that splits the number of distances into two equal parts (see **Glossary**).

[**] The mode is the most commonly occurring value.

[***] The Box-Cox transformation is widely used in the statistical literature as a tool to transform a skew distribution into a Gaussian curve. See reference[2] for a mathematical definition and references therein.

is dimensionless because the units of surface area $((distance)^2)$ multiplied by the units of the radius of gyration (R_g, *distance*) are the same as the units of volume (*distance*).[3] Thus the units in numerator and denominator cancel each other.

The expected value of *s* for a solid sphere can be shown to be 2.324. Larger values correspond to more irregularly-shaped globular objects. Thus the *s* parameter establishes a numerical scale of shape, beginning with 2.324 for the perfect sphere.

There is a clear correlation between *s* and the λ parameter of the Box-Cox transformation. The interesting thing we observed is that the observed λ values also reflect a very subtle characteristic of the irregular protein structures. The values of λ are sensitive to surface texture.

The mathematical description of the roughness or texture of a surface has been facilitated by the work on fractal geometry by the mathematician Benoit Mandelbrot (b. 1924). Mandelbrot has pioneered the use of the Hausdorff-Besicovitch's dimension to characterize the roughness of a set of points. Using some mathematical definitions, the roughness of a cluster of points can be described by numbers, with values ranging from 2.0 for a perfectly smooth surface to 3.0 for an infinitely rough surface. A "roughness" of 3.0 means that the surface is so rough that it is essentially a three-dimensional object in its own right. Numerically the texture of protein surfaces can be described with numbers ranging from 2.1 to 2.6.

This brief summary of the paper would suggest that most globular protein structures or compact domains within them (and possibly any round irregular object) could be described by three independent parameters that can be plotted on a three-dimensional Cartesian coordinate system with axes corresponding to size (the median of the distribution of inter-atomic distances, ranging from 0 to infinity), shape (2.324 and larger) and texture (2.0 to 3.0). Although size and shape are important, to me the most subtle and interesting of the three is texture.

There have been some attempts in the literature to relate surface roughness on proteins with some functional property[3] but these original seeds have not yielded any substantial fruits. In my own work, I have come to the conclusion (unpublished) that the roughness of the atomic structure of protein surfaces has two components. One is the granular texture due to the fact that protein structures are built of small spheres that we call atoms. The second is a more smoothly varying component that reflects the subtle undulations that remain after the atomic component has been removed. It is at this level that active-site cavities and inter-subunit surfaces are relevant.

With our stereoscopic vision and aided by light, our eyes can very readily perceive the properties of size and shape. Yet texture, I contend, is a property that has to be discerned by touch. How can we feel the texture of a protein molecule? Who has touched a three-dimensional model of a protein to feel its texture? One can argue that running the fingers or the palm of the hand over some of the old protein models built of mechanical parts (i.e., Kendrew brass models, see **Figure 11.1** and **Plate 13.a–d**) or plastic components (i.e., CPK models) might give you an idea of texture. Computer algorithms also exist to estimate the surface roughness of protein structures, which rapidly give you an aseptic and meaningless number. There are better ways.

In the last few years, Dr. Mike Bailey of the San Diego Supercomputer Center, A. Olson (Scripps Research Institute), and others[4] have been fabricating very attractive physical models of proteins by a process that has been described as turning paper into wood. Dr. Bailey's Laminated Object Machine (LOM) builds up a model in layers. A layer of polymer-backed paper is bonded to the growing object by running a hot iron over it. A laser cuts out the shape of the current layer, then cuts a frame around the object. This frees to paper (which is on a roll) to move forward, so the process can be repeated. When all layers have been processed, the operator applies the Michaelangelo technique—he breaks away everything that doesn't look like the object. The process can make objects up to about a cubic foot in size. They look and feel as though they are made of wood. Models of proteins manufactured by this process, typically several inches on the side, do help to appreciate the roughness of a protein surface dominated by the pea-like blobs that we call atoms and the inter-atomic recesses or cavities between them. Crevices, grooves, channels and other irregularities of the surface are very prominent in this kind of models. Nonetheless, I have a better idea.

For different materials and surfaces, first try running your hand or fingers over marble and glass. Then try the trunk of different types of trees: birches, poplars, pines, elms, maples, and oaks will all give you a sense of different roughness or textures. For the textures of different proteinaceous materials try passing your hand over the surface of sweaters made of different types of wool, such as cashmere, llama, alpaca or merino. Do the same with shirts or blouses made of silk. Finally, you can try touching the hair of your loved ones: spouses, children, lovers, and friends. After doing this, you might wonder how protein texture can result in this magic avalanche of chemical reactions in your brain and, simultaneously, a myriad of sensations and feelings in your heart.

14

The Most Delicate Proteins: Membrane Proteins

A ccording to legend, Erice, son of Venus and Neptune, founded a small town on top of a mountain more than three thousand years ago. A more historical account of the founding of Erice has been related to the fall of Troy in 1183 B.C. The historian Thucydides (ca. 500 B.C.) recounts that *"after the fall of Troy some Trojans on their escape from the Achaei arrived in Sicily on boats and as they settled near the border with the Sicanians all together they were named Elymi: their towns were Segesta and Erice."** This strategic promontory (750 meters above sea level) was fortified by the Phoenicians and contested first by the Carthaginians, then by the Romans. It was later occupied by the Saracens and reconquered by the Normans in the 11th century. A temple dedicated to the nature deity *Venus Erycina* was famous throughout the Mediterranean. Homer, Virgil, and Horace, among others, have celebrated this unique city in their poems. In spite of its ancient origin, this little gem of Mediterranean culture has a dominant medieval character with its Cyclopean walls and the remains from the old temple. In the Spring nights, walking down the streets of Erice or

---

* Paraphrased from the brochures of the Ettore Majorana Center for Scientific Culture. Courtesy of Dr. Paola Spadon and Dr. Ludovico Riva di San Severino. Also, *Encyclopaedia Britannica*, 1974 edition.

strolling by the gardens next to the old ramparts is a unique experience.

In modern times the city of Erice (province of Trapani) in the northwest of Sicily is famous for the constant pilgrimage of scientists from all over the world to its Ettore Majorana Center for Scientific Culture (EMCSC). The name of the center derives from Ettore Majorana, a Sicilian-born theoretical physicist who worked with Italian-born American physicist Enrico Fermi (1901–1954) and disappeared in 1938 on a ferry to Naples. In 1992 this institution celebrated its 30th anniversary of scientific and cultural activity. EMCSC sponsors courses in more than one hundred specialties. Participants come from universities and laboratories in more than one hundred nations around the globe. In addition, the *"Erice Statement"* written and signed in 1982 by P. A. M. Dirac, Piotr Kapitza, and Antonino Zichini, in the context of the East-West political chasm, is an important document of the scientific and moral compass of the research community. This document has been later signed and endorsed by thousands of researchers all over the world.

One of the International Schools of Crystallography held at Erice is entitled *Diffraction Methods in Molecular Biology*. It has been held every six years since 1976, initially under the scientific direction of Prof. M. G. Replacement (see Chapter 10). An amazing thing happened during this course in 1982. The audience, all seasoned crystallographers, stood up to give an spontaneous standing ovation to Hartmut Michel when he presented his diffraction pattern obtained from crystals of the photosynthetic reaction center from the bacterium *Rhodopseudomonas viridis*. As the name implies, this protein is the core of the cellular machinery that traps light photons and harnesses their energy to drive the synthesis of simple sugars (i.e., glucose) from water and carbon dioxide. Although I had the privilege of attending conferences at Erice several times, I was not present at that particular occasion; however, the event has been confirmed by different sources. You might say, weren't they all crystallographers? What is the fuss about? Don't they see thousands of diffraction patterns routinely? Yes, we do, but *that* diffraction pattern was very special; it has been reproduced on a well-known textbook on protein structure.[1]

The cellular nature of organisms implies that living systems are constituted of discrete entities that we call cells. These cells, and in fact the smaller structures within them that we call organelles, are enclosed and bounded by thin films or membranes, typically made up of two layers of lipid molecules. These membranes were originally thought to be inactive (or dormant) walls defining only the boundaries of different parts of the cell. This passive view of membranes is, one might say, completely "off the wall." Biological membranes are critical for sustaining biological activity. They actively control the exchange of matter and energy with the environment. Participating in this endless flow are protein molecules that are embedded within those membranes, and in many cases forming an integral part of them. Outside of their "natural" environment, these membrane-associated proteins are very unstable and require special treatment in their manipulation.

The astounding achievement of Harmut Michel that earned the standing ovation was learning how to extract, purify, make soluble, and eventually obtain superb crystals containing the bacterial photosynthetic reaction center. The reaction center is a complex protein aggregate that consists of four polypeptide chains, three of them named L, M and H (referring to their apparent molecular sizes: light, medium and heavy), and a fourth related to a known protein class referred to as "cytochromes." In addition, the center contains a large number of small chemical entities or pigments that absorb light of different colors, and two metals ions (Mg^{++} and Fe^{++}). The crystal structure was presented only a few years later by Deisenhofer, Michel, and Huber.[13] It revealed how the protein subunits (L, M, and H) provide the scaffold for the binding of all the other components, and how this functional unit allows the flow of electrons across the two sides of the membrane. This three-dimensional structure was the first high-resolution structure obtained from an intact membrane-bound protein; it occupies a very special place in the heart of structural biologists. **Plate 14.1** presents the "classic" cartoon drawing of the photosynthetic reaction center of *R. viridis* as displayed in 1985 (Courtesy of Hans Deisenhofer).

Crystallization of membrane-bound proteins is a specialty by itself and requires sound knowledge of biochemistry and physical chemistry as well as a special crystallization "green thumb." The basic ingredient for the successful handling of membrane proteins is the use of small molecules containing polar and non-polar groups at the same time but on different sides of the structure. These molecules are called detergents, and in a manner similar to our household detergents, help the stability, manipulation and extraction of oil-liking entities. The insight of Harmut Michel was to use detergents to mask the hydrophobic (water-hating) surfaces of the photosynthetic reaction center, associated with the cell membrane, and make them more hydrophilic (water liking). Although the concept now is well-established and there are established protocols to manipulate these proteins, only a limited number of high-resolution structures of membrane proteins are available.

You may say, if they are so difficult to handle, why do we have to bother? Don't crystallographers have enough work with the more soluble, well-behaved proteins? Alas! That is easy to say, but unfortunately many of the proteins performing the most subtle and exquisite tasks within the organisms are membrane-associated proteins. If living systems are interconnected mosaics of cells, it follows that communication among the cells, within the cells and with the external surroundings is of pivotal importance. Broadly speaking, all these functions are carried out by membrane-bound proteins, described with names such as gap-junctions, ion-channel proteins (for instance, Na^+ or Ca^{++} channels), and receptor proteins of different kinds. Among receptor proteins, the ones associated with our sensory organs are of special interest to us, being as we are sensual creatures. I will not try to praise or expound on the wonders of the universe of our sensory organs here. My words will be dispersed into the air by the gentle zephyr of Diane Ackerman's lyrical prose in her exquisite, delicate and evocative book about the senses entitled *The Natural History of the Senses*.[2] I will only attempt to illustrate with one example how the atomic machinery of those senses is connected with the design and artistry of protein structure.

From a physiological viewpoint, there are many sensory modalities: chemical, which could include taste and smell; somatosensory, referring to the detection of signals such as touch, pressure, temperature and pain; kinesthesia, encompassing the sensations having to do our own motion; balance or equilibrium, telling us where we are; and the ones we are most familiar with, hearing and vision. In spite of this wide variety of signals, the sensory stimuli are only of three main types: chemical, mechanical, and electromagnetic (or light related). All receptors can be classified according to the type of energy input they receive. Thus broadly speaking there are chemoreceptors, mechanoreceptors, and photoreceptors.[3]

I would like to focus the discussion in the sensory processes related to chemical detection and particularly smell, even though we do not have a complete molecular mechanism to explain all the necessary steps. From an evolutionary standpoint, it seems as though the first single-cell organisms, once they could be defined as such, had to be able to detect molecules and move either to or away from them. Thus chemical detection is probably the most ancient and primitive of all sensory systems. For we humans the sense of smell might appear as if it has lost its essential nature, but I would like to bring it to the forefront as a critical part of our sensorial universe. In our current world, dominated by visual images and recorded sounds, which can be manipulated in so many innumerable ways, the sense of smell carries with it (at least for the moment) the connotation of proximity and authenticity. I often walk on the shores of Lake Michigan and indeed it looks like the sea; when the surf is in action, it even sounds like the sea. However, it does not *smell* like the sea, and certainly not like my favorite ambience, the unique salty aroma of the seaport in Donostia (San Sebastian, Spain) mixed with the scent of fresh, grilled sardines over an open wood fire. Possibly, and I hope I won't live to smell it, when the Internet and our movie screens might also transmit and convey odors, then all our experiences would be transmutable and vulnerable. Until then, I hold on to the sense of smell to extract the essence of my personal experiences. A quotation from Pierre Perret, a French specialist in wild mushrooms, which I found in one of those generic airline magazines put it in a way that I can only reproduce here:[4]

"Ah l'odeur d'un champignon frais. Seule celle d'une femme qu'on aime peut se comparer a une telle merveille."

"Ah, the aroma of a wild mushroom. Only the scent of the woman one loves could compare with such a wonder."

Either of those two exquisite sensations is produced by a cascade of events at the molecular level, many of which have to do with membrane-associated proteins. Consider just for a moment the molecular events that have to take place for you to sense that unique aroma. A small concentration of various molecules, in the jargon of physiology called "odogens" (i.e., that generate odor) will be gently blown inside your nose and onto the olfactory epithelium located in the dorsal recess of the nasal cavity. In the mature olfactory-receptor neurons, there are long cilia that absorb and direct the odogens to the receptor molecules in the cell membrane. The minute molecular signal of a few molecules is recognized by the receptor protein in a lock-and-key fit, resulting in a signal being transferred across the membrane. The receptor may have a structurally constant region to transmit this signal, but it will also have a more "variable" region containing the shapes that will fit onto the different stimulating molecules or odorants. It is known that the signal is amplified by G-proteins, so named because they act on GTP, a relative of ATP, the main energy currency of the cell. G-proteins activate a chemical reaction catalyzed by an enzyme called adenylate cyclase (AC), which synthesizes a second messenger in the cascade called cyclic AMP (cAMP). It is this second messenger, generated inside the cell in response to the odorant outside the cell, that alters the state of an ion-channel protein and generates the receptor potential that produces the nerve impulses traveling to the brain. Certain odor molecules might also bind directly to the ion-channel protein and produce a direct nerve impulse. At the other end, these nerve impulses have to reach a certain threshold to be detected, their magnitude needs to be estimated, analyzed, discriminated and finally recognized, eliciting the appropriate response, which is another molecular prodigy by itself.

I did not recite this concise series of events to impress you. They can be found in exquisite detail in any physiology or neurobiology textbook. The purpose was just to give you an overview of the molecular events occurring literally under our skin, to be able to perceive a sensorial input related to the sense of smell. The molecular components are somewhat similar for vision; both share a mechanism referred to as G-coupled receptors. We do not yet have a complete three-dimensional image of all the proteins involved in any of the sensory cascades. We have bits and pieces of the different ones obtained from various organisms, using different tricks to purify certain fragments of the proteins involved and to crystallize them. We have a high-resolution image of the receptor protein for light, called rhodopsin, from bovine rods.[5] The crystallographic work that permitted the elucidation of the structure of rhodopsin had to be carried out in the dark.[5] It is a prototypical membrane protein, containing seven helical rods spanning the entire width of the membrane, which cradles a small molecule that changes structure upon capturing a photon of light (**Plate 14.2**). There are also several structures of the G-proteins bound to various intermediates of the reaction (i.e., GTP, GDP). Recently, the structure of a receptor molecule that channels potassium ions (K^+) was also unveiled, displaying a funnel-like shape that surrounds a narrow passage that acts as a filter for the small potassium ions.[6] The unveiling of the three-dimensional structure of these and several other membrane proteins are remarkable technical accomplishments in the field that have been published in the most influential scientific journals. The crystallographic results have stirred tremendous scientific excitement in fields of research totally unrelated to crystallography such as physiology and vision research.

Certain protein crystallographers have worked very hard, and have been fortunate enough to unveil the structures of other membrane proteins playing just as essential, and just as subtle, molecular functions as the ones involved in sensory perception. Cytochrome c oxidase,[7] an enzyme involved in the last steps of the oxidation (removal of electrons) of our foodstuffs to simpler components such as carbon dioxide and water, is a large multi-protein complex. Adenosine Tri-Phosphate synthetase, another integral membrane protein, is a beautifully intricate molecular machine. ATP synthetase generates ATP, the essential energy currency of the cell, exploiting the charge and concentration differential (gradient) between the outside and the inside of the cell.[8,9] These charge and concentrations gradients are produced and maintained by other membrane-associated proteins and by the structural integrity of the membrane itself. Yet another example is an amazing pore-like protein named "porin" with a structure resembling atomic wicker work, forming a basket-like tunnel traversing the membrane and acting like a pore.[10] **Plate 14.3** presents a collage of these intriguing and fascinating protein structures. Many other ones are being worked on in laboratories all over the world. We can confidently expect them to see the light of day in the near future.

As impressive as these findings may be, they are only the beginning. It has been estimated that over one thousand G-coupled receptors exist in high vertebrates, and amazingly enough, of those more than five hundred will be odorant receptors. As you know, I have skipped over the other sensory pathways and not discussed other communication networks in the cell, with mechanisms just as subtle, intriguing, and interesting. In all of them there will be a myriad of proteins associated with cell membranes and ion-channels, whose function will have implications for the understanding of many human pathologies or "channelopathies."[11] I may only hope that I have whetted your appetite, since the study of the molecular mechanisms involved in what is called "signal transduction" is a vibrant and exciting area of study. The macroscopic levels of observation are physiological recordings of ion channels, cell-clamp experiments and synaptic potentials among others. At the molecular level the actors are proteins such as receptors, ion-channels, G-proteins and small chemical entities known as neurotransmitters, hormones, odorants, and others. These atomic players are constantly in action to allow us to see, hear, smell, touch, and move: in short, to be alive. Given the difficulties faced in handling membrane proteins, I think that you can now fully appreciate two important facets of the field of membrane proteins. On one side, we have the excitement of the crystallographic community in contemplating the diffraction pattern that Harmut Michel obtained from a membrane protein that represented a milestone in the field. On the other side, we have the enormous challenges that await structural biologists working with these most delicate and versatile structures.

VI

new technologies

15

A Brilliant Star in the Midwest Illuminates the Future of Macromolecular Crystallography*

"Molecular anatomy will be the foundation of medicine in the 21st century as was human anatomy five centuries earlier [...] In as much as synchrotron radiation is the primary means by which large scale biological structural information will be obtained in the future, continued support is of the utmost importance."

Arthur Kornberg

The Advanced Photon Source (APS) located west of Chicago at Argonne National Laboratory, in the heart of the American Midwest, is the first third-generation synchrotron source in the hard X-ray range in the U.S.A. It was dedicated on May 1, 1996. This milestone marked the beginning of a bright future for the community of

---

* Adapted from the version published in *Physics News in 1996* to commemorate the commissioning of the APS. "A Brilliant Star in the Midwest Illuminates the Future of Crystallography" (AIP, New York, 1997, P.F. Schewe and B.P. Stein, Eds, pp. 36–39). This chapter is somewhat more technical to impress upon the reader the technological resources available at synchrotrons and the corresponding advances that follow from the use of these unique X-ray sources. It will be impossible to make fully comprehensible all these technologies in a brief article. Arthur Kornberg's quotation obtained from the Executive Summary of the Report of the Basic Energy Sciences Advisory Committee Panel on Synchrotron Sources and Science, Deparment of Energy, unpublished report, 1997).

macromolecular crystallographers in the U.S. and for the structural and crystallographic community in the world at large (**Plate 15.1**).

Synchrotrons are devices that accelerate and retain charged particles within circular orbits; synchrotron radiation is the electromagnetic radiation given off tangential to the ring as the particles are confined to their enclosed orbits. Critical characteristics of synchrotron radiation for macromolecular crystallography[1] are high brilliance, tunability, and the time-resolved structure of the beam. Brilliance is an indication of the intensity (photons/second) of the beam as seen on a unit of area, unit of angular tilt and within a narrow band (0.1%) of wavelength acceptance. High brilliance permits collection of high quality data from small, weakly diffracting crystals. Tunability is the ability of "dialing in" the wavelength of the X-rays. Conceptually, this is analogous to selecting a color in the visible spectrum or, in our metaphor, changing the "color of the wind." This capability is critical to perturb the electronic structure of certain heavy atoms, which facilitates dramatically the solution of the protein structure. This is done by collecting data at different wavelengths and combining the observations by the method of **M**ultiwavelength **A**nomalous **D**iffraction[2] (**MAD**, see Chapter 9 *"The Combs of the Wind"*). The time-resolved nature of the beam can be utilized to analyze enzymatic or molecular processes at the millisecond to microsecond time scale. In the third generation synchrotrons it is common to alternate two types of stations within what is defined as a "sector." One, referred to as "bending **magnet**" (**BM**-lines), utilizes the X-rays produced by the electrons in the storage ring as they bend in a curved trajectory. Next, straight sections within the ring perimeter allow the placement of "insertion **d**evices" (**ID**-lines), which are a complex array of magnets that enhance, collimate, and magnify the X-rays emitted by the traveling electrons. Pulses of X-rays can be generated at synchrotrons as bunches of electrons travel around the ring (approximately one mile in circumference at the **APS**), at a speed close to the speed of light (300,000 Km/second).

Not surprisingly, the external appearance of many synchrotrons is that of a gigantic ring that can measure over one mile in circumference for the largest ones. Inside, one walks within a concrete building with imposing massive walls that enclose the beam of electrons forced into a closed trajectory by the action of potent magnets. Initially the space outside those walls, which is referred to as the experimental hall, is empty and is analogous to a large circular tunnel. In time, however, this is the space that scientists of many different specialties fill with steel beam enclosures, sophisticated instrumentation and large lead hutches crowded with electronic equipment. All those components make up the experimental stations or beamlines, each one with a specific experimental design to bring the X-rays emanating from the circulating electrons to a certain experimental sample, which may be a surface, a crystal, or a fiber (**Plate 15.2**). Quite often as I walk around the ring the memory of my father and I walking around the different chapels of a magnificent cathedral comes to my mind. The synchrotron is a unique cathedral dedicated to technical ingenuity, scientific inquiry, and devoted to research rather than a temple of contemplation, worship and prayer.

It is often not realized that the dramatic increase in the number and complexity of the macromolecular structures determined in the last decade or so is, in no a small measure, due to the availability of tunable X-ray sources of high brilliance. It is in this framework that the dedication of the **APS** marked a new inflection point in the development of macromolecular crystallography. If one considers the impact that the discovery of X-rays by Wilhem Conrad Roentgen over 100 years ago has had on our knowledge of the material world that surround us, then one realizes the importance of this unique form of radiation. Since its discovery, the brilliance of available X-ray sources has increased by a trillion-fold (10^{12}), and the impact that these sources will have on our understanding of the material world can only be a matter of conjecture.

The first protein crystallography stations using synchrotron radiation were operating in the parasitic mode, utilizing the radiation given off by the synchrotrons designed for experiments in high-energy physics. These first-generation-synchrotrons were providing approximately 10,000-fold more brilliance than the conventional in-house sources of the time. From this early usage a new generation of synchrotron sources came on-line in the early 1980s, specifically designed to produce more intense X-ray beams and achieving an additional 10,000-fold in brilliance (Daresbury Synchrotron Radiation Source, SRS, England; National Synchrotron Light Source (NSLS, Upton, NY; Photon Factory, Tsukuba, Japan).

There are two general considerations that will attest to the future importance of third-generation synchrotron X-ray sources like **ALS** (Berkeley, U.S.A), **ESRF** (Grenoble, France), **APS** (Chicago, U.S.A), **SPring8** (Hyogo, Japan), *Elettra* (Trieste, Italy), and **CLS** (currently under construction in Saskatoon, Canada). On the technical side, these facilities have been designed specifically to generate tunable X-ray beams of unprecedented brilliance (an additional 10,000-fold over second generation sources). On the sociological side, these installations have been planned and executed with the user community in mind. It is estimated that 2,000 researchers will visit the **APS** per year, with an approximate floating population of 300–400 investigators at any one time. The planning team of the **APS** has provided additional space on the periphery of the ring for offices and laboratories (laboratory office modules or **LOM**s) adjacent to the experimental hall. These clusters of five triangular extensions give the **APS** its unique serrated appearance. A user residence facility located within walking distance from the storage ring provides accommodations for up to 240 users and computer links to the experimental hall (**Plates 15.1** and **15.2**).

For experiments in the realm of macromolecular structure, four Collaborative Access Teams or **CAT**s are currently in operation at stations with different but complementary interests. **Bio-CARS (sector 14)**, within the Consortium for Advanced Radiation Sources, currently operates three experimental stations designed for crystals with very large unit cells such as crystalline samples of viruses and ribosome-like aggregates. The stations allow data to be collected in both monochromatic and polychromatic mode including **MAD** phasing experiments. The structure of the bacteriophage HK97, a type of bacteria-eating virus containing double-stranded DNA, has been mapped at 3.5 Å by the group of Prof. J. E. Johnson of the Scripps Research Institute (California) using more than twenty-one million reflections measured on one of these beamlines.

The Structural Biology Center (**SBC-CAT, sector 19**) has been conceived as a user facility for macromolecular crystallographers. The technical facilities have been specifically designed to support (i) very fast **MAD** phasing experiments; (ii) data collection from microcrystals; (iii) data collection from crystals of large macromolecular aggregates; and (iv) routine rapid data collection from similar or related crystals. The superb quality of the data collected at this facility has been of critical importance for the high resolution structure determination of the 50S and 30S ribosomal subunits (see below).

The Industrial Macromolecular Crystallography Association (**IMCA**, a consortium of twelve chemical and pharmaceutical companies) together with the Illinois Institute of Technology (**IIT**, Chicago, IL), form the **IMCA-CAT (sector 17)**. This group is currently operating two beamlines (one sector) with the objective of collecting diffraction data from macromolecular crystals of biomedical interest to aid in their drug-design and protein engineering programs. Routine access to this resource is having a tremendous impact in the design of novel pharmaceuticals. In addition, researchers from the Center for Advanced Research in Biotechnology (Rockville, Maryland), the institute for Genomic Research (also in Rockville), and the IIT use the **IMCA-CAT** facilities to solve the structures of proteins of unknown function from the agent responsible for the flu (*Haemophilus influenzae*). The successful operation of **IMCA-CAT** is a tangible example of how pharmaceutical companies can share and benefit from a large experimental facility built by the U.S. government, using the taxpayers' money. Although the U.S. government built the **APS** subsidized by public funds, the **IMCA** consortium paid for the planning, design, and construction of the beamlines. In addition, the consortium pays for the current operation of the beamlines and also hourly charges for the proprietary usage of the facility. The data gathered at routine crystallographic experiments performed at the beamlines designed, constructed and financed by the **IMCA** consortium are critical for the rapid design of novel pharmacological entities that will eventually improve the quality of our lives.

Bio-CAT has one experimental station (**sector 18-ID**) to study the structure and dynamics of partially ordered samples such as membranes and fibers. The equipment, software and laboratories of the **Bio-CAT** station are available to a wide variety of users, predominantly from academic institutions. Recently this facility has been used to obtain structural information on the dynamics of thoracic muscle contraction in living samples of the fruit fly *Drosophila melanogaster*. The other station in the sector (**18-BM**) is being designed for a novel imaging technique called **D**iffraction **E**nhanced **I**maging (**DEI**), which could have tremendous impact in the early detection of breast cancers. Another sector (**sector 22**) is currently being developed to accommodate a consortium of academic institutions within the geographical region of the South Eastern portion of the United States (**SER-CAT**), which will focus on macromolecular crystallography and structural biology.

There are many more experimental stations around the **APS** synchrotron ring to a total of thirty-four sectors. Besides macromolecular crystallography, the radiation emanating from the ring is used for the study of polymers and other materials, environmental migration and degradation of pollutants (**Geo-CARS, sector 13**), surface science and catalysts and many other important materials. A glimpse of the research activity of the **APS** can be obtained from its recent publications and its Internet web site. A schematic of the disposition of the different **C**ollaborative **A**ccess **T**eams (**CATs**) around the ring is presented in **Plate 15.2**, grouped by sector and discipline, including the beamlines currently under construction.

Macromolecular crystallography has reached a stage where every year several thousand protein and nucleic acid structures are routinely reported and deposited into the Protein Data Bank (PDB), which currently (May 2002) holds over 18,000 entries. This macromolecular repository celebrated its 25th year of existence 1998 and is currently a critical resource for structural biologist all around world. With the advent of massive structure determination efforts within the framework of structural genomics, the PDB can only grow and develop tools to expedite the dissemination of structural information.

Within the multitude of lectures, presentations and posters (2250 abstracts from 65 countries) at the XVII Congress and General Assembly of the International Union of Crystallography (IUCr, Seattle, August, 8–17, 1996), outstanding crystallographic results were presented and discussed. The reported structures had tremendous impact in a wide variety of biological areas: cell cycle regulation (cyclin) and cell death (Bcl-XL);[3] energy production (Cytochrome c oxidase);[4] catalytic hammerhead ribozymes[5,6] (enzymes made up of RNA instead of protein); eukaryotic transcription complexes;[7] protein folding (chaperonins)[8] and the action of cell toxins, as well as many more. Just as novel and exciting results are expected at the XIX assembly of the International Union of Crystallography (IUCr2002) in Geneva, Switzerland in the Summer of 2002.

A striking example of the impact of synchrotron radiation in elucidating the structure of complex aggregates of biological macromolecules was the work of the late Prof. Paul Sigler (1934–2000).[7] He and his collaborators were able to unravel the structure of a ternary complex containing: a piece of DNA called the TATA promoter; a transcription factor TFIIA; and the TATA binding protein TBP. This complex is a critical component of the machinery in the cell that translates the genetic information contained in DNA into the proteins that are the majority of the chemically active parts of organisms. The structure was solved by synchrotron radiation tuned to two different heavy atoms in the structure, selenium in the protein part and bromine in the DNA. The macromolecular aggregate presents a bundle of four helices that extends away perpendicularly from the TBP/TATA complex and allows for further recognition by other participants in the process of the protein production in cells.

All of these successes have been crowned just at the end of the millennium by the first images at medium resolution (between 4 and 8 Å) of the key components of the molecular machinery of protein synthesis, the ribosome.[9,10] Characterized in the sixties as discrete particles with specific sedimentation properties, ribosomes are large, nonsymmetrical macromolecular aggregates that contain protein and RNA. Ribosomes are composed of two subunits. The completely assembled ribosome is characterized by a sedimentation coefficient of 70S and is composed of two unequal subunits. The smaller one has a sedimentation coefficient of 30S and the larger one has a sedimentation coefficient of 50S. The pioneering work and unfailing vision of Ada Yonath and her colleagues (Weizman Institute of Science, Rehovot, Israel, and the late H. G. Wittmann, University of Berlin) made it possible to grow resilient crystals of ribosomes from microbes from the Dead Sea (*H. marismortui*). Moreover, she showed that by freezing the crystals with techniques developed by H. Hope (University of California, Davis) one could obtain good data at synchrotron sources all over the world, leading the way to the next stage of structural work on the ribosomes. In 1999, we have seen the publication of the structures of the small subunit (Ramakrishann and co-workers, 5.5 Å);[11] the large subunit (P. Moore, T. Steiz, 5 Å);[12] and the first glimpses of the fully assembled particle (H. Noller, M. Yusupov, and G. Yusupova).[13] At the 50th annual meeting of the ACA in St. Paul, Minnesota, the groups of T. Steiz and P. Moore reported the structure of the large (50S) ribosomal subunit at 2.4 Å resolution. Amazingly enough, the structure shows a gigantic piece of catalytic machinery made predominantly of RNA; the fine detail was made possible by the use of radiation from third generation synchrotrons.[14] At the time of this writing, the structure of the 30S subunit has also being been reported by two groups,[15,16] and the modes of interaction of antibiotics with this piece of cellular machinery are being scrutinized at the atomic level.[17] Macromolecular crystallography combined with the wide use of synchrotron radiation has reached the zenith as the most powerful technique to produce atomic images of the molecules that make life possible. Likely, new techniques using the brightness, availability, and versatility of synchrotron radiation or free electron lasers (FEL) will open new horizons on ways to visualize molecular structure in a not too distant future.

One more example of experimental wizardry is possible with synchrotrons that I have to mention because it may be very important in the future to study the dynamics of biological processes. I will remind you that in the original diffraction experiment by Max von Laue and collaborators, the X-rays illuminating the sample were multicolored, they had many different wavelengths. The same thing is true for the raw or "untreated" radiation produced by synchrotrons. Protein crystallographers normally select one particular wavelength for their experiments, although this specific wavelength can be changed as in MAD experiments. However, given the enormous intensity of the X-rays coming from synchrotrons, illuminating the protein samples "à la Laue" permits the collection of entire data sets in a very short time (milliseconds to microseconds, 10^{-3} to 10^{-6} seconds). This experimental capability combined with pulsed nature of synchrotron radiation offers unique possibilities for the future time-dependent study of catalytic processes in biology.[18]

A truly brilliant star has been recently born and shines in the Midwest Sky of the North American continent. Together with the ones already existing in the world and the ones currently under construction, they will all illuminate our path towards a better understanding of the intricacies and subtleties within the atomic landscapes, islands, continents, and universes that surround us, and of which we are an integral part.

16

The Lunar Element and Our Collective MADness

If you have reached this point, you probably have noticed that the city of Grenoble (Department of Isère and capital of the Dauphiné province) in France has had a historical connection with crystallography. It was in this city that J. B. J. Fourier began to work systematically on his *memoire* on the analytical theory of heat, which had an enormous influence on the mathematical theory of crystallography. Quite appropriately, the university at Grenoble is named after him. Nowadays, Grenoble is also famous among crystallographers because of the presence of an outstanding example of synchrotron sources: the European Synchrotron Radiation Facility (**ESRF**) is located there. Consequently, there is a constant flow of crystallographers and other scientists using X-rays that benefit from the superb characteristics of this source of X-rays. I would like to single out one among those itinerant scientists, Prof. Wayne A. Hendrickson, currently Prof. of Biochemistry and Molecular Biophysics at Columbia University College of Physicians and Surgeons and investigator of the Howard Hughes Medical Institute. He did his doctoral work in biophysics with W. Love at Johns Hopkins and later went to the Naval Research Laboratory (NRL) for postdoctoral studies with Jerome Karle, a prominent theoretical crystallographer. His strong interest in the development of crystallographic techniques that exploit the "anomalous" scattering of atoms probably stems from his interactions with Dr. Karle.

To appreciate his achievements, let us recapitulate what is involved in the process of obtaining the three-dimensional structure of a protein as summarized in **Plate 15.1**. Good quality crystals are exposed to an intense source of X-rays and data collected for a native, "unperturbed," crystal. Atoms with a large number of electrons (i.e., heavy atoms, such as platinum (Pt), mercury (Hg), and gold (Au)) are soaked into the protein crystals and data are collected for at least two additional derivatives (**MIR** method). The presence of these heavy atoms possibly disturbs the crystal lattice and introduces errors. However, in the past it was the method of choice to calculate the phases of the Fourier terms, which allows the reconstruction of the molecules in the crystal. Because of the singular structure of their surrounding electronic clouds, some of these heavy atoms are very sensitive to the "color" (i.e., wavelength) of the X-rays, and strong anomalous effects had been used to expedite and improve the overall process if tunable sources, like synchrotrons, were available to collect data. By combining the flexibility of synchrotron sources with the use selenium (Se) to replace sulfur in protein structures, Prof. Hendrickson has pioneered a method that has proven to be rapid, efficient, and effective in solving structures of biological macromolecules. The lines that follow are meant to express my admiration and respect for his insight and achievements. I am certain that many of my colleagues will join me in this modest homage.

In the old part of Grenoble there is a café named *Café de la Table Ronde*, right across from the old Palais de Justice in the *Place de St. André*. According to some of the newspaper clippings hanging from its rustic walls, this coffee house was the second in France, founded in 1739, seventy years after the first coffee house in France (the *Café d'Europe in Paris*). Among its clientele this café boasted actors such as Fernandel and Gerard Depardieu; folksingers and balladeers such as Jacques Brel and George Brassens; and writers of the caliber of Pierre Choderlos de Laclos (1741–1803; *Les Liaisons Dangereuses*, object of an American film by the same title) and Stendhal (1783–1842;

pseudonym of Marie-Henri Beyle, *Le Rouge et le Noir*), who was a native of Grenoble. I am certain that in most recent years it has also been the place of gathering for many of the most illustrious scientists that visit the **ESRF** in Grenoble every year. However, they were not listed yet.

I was sitting at this café late in the evening on June 16, 1996 at the conclusion of the workshop on **M**ultiwavelength **A**nomalous **D**iffraction (**MAD**, Grenoble, June 10–16, 1996) making time until the departure of my TGV train back to Paris. Somehow, a photographer managed to capture the very same spot at which I was sitting in one of the Grenoble images (**Plate 16.1**). The square where the outdoor tables were set was full of life, and a full moon was shining up on one of the corners of the quadrilateral defined by the roofs of the buildings around me, including the parish of *Saint André*. As I remember that night, and the subsequent developments in our laboratory, I feel as if, in that particular moment, I was "moonstruck" in a scientific rather than romantic manner. At that time, our laboratory had just solved, in a matter of only a few weeks, its first structure using the MAD methodology. From then on I have felt captivated by the "power and sheer elegance of the method"[1] derived from the substitution of sulfur by selenium in the protein (you may like to refer back to Chapter 9 *"The Combs of the Wind"*).

Selenium (from Greek *selene*, "moon") derives its name from its silvery appearance and was recognized as an element in 1818 by Jöns Jacob Berzelius. Its location in *The Periodic Kingdom** is in the Eastern Rectangle (the p-block), right after the isthmus that connects the strong metals (Western Rectangle) with the nonmetallic elements. Yellow Sulfur is its northern neighbor. Its gray metallic, lunar-like appearance was probably the main reason behind its name, but its ability to exist in several different colored forms, akin to the phases of the Moon, established for me another connection with our night planet. As an amorphous (noncrystalline) powder it is red but trans-

---

* The language refers to the regions of the Periodic Table, central icon to the classification of the chemical elements as described in reference.[2]

forms easily into a black vitreous glass. As a crystal it can be red or gray, with the latter being the most stable under ordinary conditions. By itself it is not poisonous but many of its compounds are very toxic. However, it is also an essential mineral that has to be provided in the diet of certain animals.

After that night, it seems as if I am under the spell of this gray metalloid element and I think that I am not alone in view of the many structures that are solved nowadays using this methodology. The spell of this lunar element on protein crystallographers is undoubtedly due to the unique properties of its electron cloud with an absorption K-edge at 12.658 KeV. This is an "anomalous" discontinuity of response at an energy of 12.658 Kilo-electron-Volts, which corresponds to a wavelength of the X-rays of 0.9795 Å, well within the range of synchrotron sources. Its strong attractive influence is also due to the fact that it can be easily incorporated into our protein samples and that most of the time our crystals do not even notice it[3,4] (see Chapters 9 "*The Combs of the Wind*" and 21 "*The Soft Engineers*").

As judged by the number of protein structures solved by this method in 1997,[5] my lunar hallucinations have spread to a large number of members of the protein crystallography community and may soon reach the level of a collective MADness. The origins of this hysteria can be traced back for a number of years and I will not elaborate on this.[6,7] However, I would argue that there must be an instigator, a culprit, for such a large group of "intelligent" people to become engulfed in such a collective state of agitation. I have not taken a poll but I think that, for once, the community is unanimous and unambiguous as to whom that person is, and I will not even have to name the author of such witchcraft.

May these few lines be an expression of respect and admiration for the leader of this unusual and benign cult. May these words pay homage to one of the unsung heroes of the structural revolution in Biology, who has been justly honored with many awards within and outside the crystallographic community. May these letters express my personal gratitude, that of my colleagues, and of the community at large, to such an "anomalous" and visionary colleague. He has opened for us a path to solve new structures, illuminated by the pale light and color of the lunar element combined with the brightness and tunability of our most brilliant X-ray sources. When I go to the *Café de la Table Ronde* next time, I would like to see the names of some scientists among its illustrious clientele, and our MADman among them.

17

On Gold Rings and Synchrotron Rings

Dedicated to the people who planned,
designed, built, and currently work
at synchrotrons

In a previous chapter I have written about the technical specifications of synchrotron rings and about the immense possibilities that the power and brilliance of these third-generation sources will offer for the investigation of the atomic universes that surround us. This time, I would like to take a different, more philosophical, viewpoint.

Rings are very special artifacts and icons in Western culture and most likely in many other societies also. Gold rings or rings of other precious metals, engraved and decorated with precious stones, have always been the attributes of kings, queens, and other high dignitaries. Quite often throughout history their owners may claim special privileges of power, property, or influence in both religious and secular organizations. Additionally, they are commonly associated with material wealth and used as tokens of special liaisons, alliances, or compromises.

Among human beings, the most special and at the same time the most common are the agreements of engagement or marriage. Rings are exchanged between the bride (or the bride-to-be) and the groom (or the groom-to-be) as a symbol of eternal fidelity and unending love. I will only mention the *"claddagh"* ring in the Irish culture, a heritage from their Gaelic ancestors as a symbol of friendship, loyalty, and love, also used as a betrothal ring. The *claddagh* presents a prominent heart in the middle of the ring, which, by the way it is pointing, indicates whether the carrier of the ring is free or engaged to another person.

In the realm of dramatic musical arts, there is a monumental composition, which unfolds around a gold ring—collectively referred to as "*The Ring*" or "*The Ring Cycle*." I am referring to *Dem Ring des Nibelungen* by Richard Wagner (1813–1883), arguably the most massive, complex, and expressive dramatic music of the Western culture.

In its final form, *The Ring Cycle* consists of the introduction (a major opera) *Das Rheingold* (*The Gold of the Rhine*) followed by three complete operas entitled: *Die Walküre*, *Siegfried*, and *Götterdämmerung* (*The Valkyrie, Siegfried*, and *The Twilight of the Gods*). The Ring represents the culmination of the complete work of art ("*Gesamtkünstwerk*") conceived by Richard Wagner as the perfect enmeshing of music, words, and action. This composition is arguably the most ambitious, most complex, and most expressive of the musical drama.

The text (I would not dare to call it libretto) of *The Ring Cycle* originates from the research that Wagner did on the mythology and sagas of the Teutonic and Norse cultures circa 1849. The most important characters are a god (Wotan) and a goddess (Fricka), who would like to live in a magnificent and impregnable palace in heaven (Valhalla) (**Plate 17.1**); a hero (Siegfried); a disobedient heroine (Brünnhilde) who turns into a sleeping maiden atop a fire-girth rock; villainous, shrew dwarfs (Alberich and Mime) who mold the magic ring and forge a magic helm (Tarnhelm); and strong, brutish giants (Fafner and Fasolt) who agree to build Wotan's fortress in exchange for Freia (the goddess of beauty), and one of which later transforms into a dragon to guard the gold of the Rhine.

From the first literary sketches to the time that the last note was written, it took 26 years to complete *The Ring Cycle*. This includes a long pause of twelve years, during which Wagner broke off the composition of *The Ring* (in the middle of Act II of *Siegfried*) and concentrated on *Tristan und Isolda* and *Die Meistersinger*. When *The*

Ring was resumed in 1869 and completed in 1874, Wagner had evolved considerably, both musically and philosophically, and the work did not have the concise logic that guided its original gestation. In addition, Wagner had written to his imprisoned revolutionary friend, August Röckel, of the 1849 Dresden[1] uprising:

"I believe that it was a true instinct that led me to guard against excessive eagerness to make things plain, for I have learned to feel that to make one's intentions too obvious risks impairing the proper understanding of the work in question; in drama—as in any work of art—it is a question of making an impression, not by parading one's opinions, but by setting forth what is instinctive."

All these factors compound to make the full, complete, logical understanding of *The Ring* almost impossible, and certainly beyond the scope of these lines. The first complete public performance of *The Ring* took place in Bayreuth, directed by Hans Richter in August 1876, a quarter of a century after it was first conceived. Overall, we can take *The Ring* tetralogy as an "allegory and a conspectus of human relations and society—possibly also, as a testament of a disillusioned would—be reformer."[2]

Yet, what is *The Ring* about? In spite of the disagreements concerning the meaning of *The Ring*, most people will agree that it is about the conflict between love and power. In the opening scene of *Des Rheingold*, Wellgunde, one of the Rhinemaids, says:

"The inheritance of the world would be won by him who made The Ring from the Rhinegold, it would vouchsafe him limitless power."

This was followed immediately by the chilling strings and the announcement by Woglinde, the other Rhinemaid:

"Only he who forswears love...."

Immediately after, the chief villain, Alberich, tears the gold from the rock, renounces love as a prerequisite to the power and possession of *The Ring*, and henceforth gets the action of the entire drama underway. The tension increases when Alberich, dispossessed of the gold of the Rhine, *The Ring* that he was able to craft from it, and the magic helmet by Wotan, curses *The Ring* to be the cause of death to whoever owns it. What Wagner means or implies by love during the duration of *The Ring Cycle* changes; yet, from a simple viewpoint, it seems as if Wotan and Alberich provide the essential dynamic force of the drama by renouncing love for power. In fact, Wotan, this very human god, says to Brünnhilde at the beginning of his monologue in Act II of *Der Walküre*:

> "When the joy of young love departed from me, my spirit longed for power."

Back then to synchrotron rings. We have amassed the "gold of the Rhine" and from it we have built not one, but many synchrotron rings around the earth. These rings have given us a tremendous amount of power to pierce and penetrate matter and to uncover many of its well hidden secrets (**Plate 17.1**). We can probe the material universe at the atomic level within the animate and inanimate domains. We have gathered, and we will continue to gather, enormous amounts of data, which will translate into new forms of knowledge. We know well that knowledge is power. Do we have to renounce love and beauty?

The answer should be a resounding "No!"

We should continue to illuminate our samples with evermost brilliant sources of radiation and use all possible ranges of the electromagnetic spectrum as well as all possible experimental configurations. However, we should also retain the admiration and awe for the cosmos around us. We must convey a sense of beauty for what we discover and respect for the "mysteries" that are still the undiscovered facets of this multisided microcosm. We should consider our new structures not as climbing rungs in our career ladders but as prodigious pebbles—precious jewels—that we have been fortunate enough to unveil and privileged enough to present to humankind. We have to retain the joy and enthusiasm of young love toward the puzzles that still have to be discovered within the current paradigms of science. In addition, we should cultivate an insatiable thirst for the "unknown unknowns"[3,4] that exist both inside ourselves and outside in our surrounding universes. Only in this way will the immense power of our Synchrotron Rings not destroy us.

18

Cathedrals and Synchrotrons for the 21st Century

Dedicated to the people of Dresden, Germany

We were walking in the *Place d'Armes of vieux* Montreal outside the *Notre Dame* Basilica after a brief visit inside. Although not very grandiose, this cathedral has harmonious architecture and the light penetrating through beautiful stained glass windows illuminated its interior and conveyed a peaceful and serene ambiance. The first and original chapel was a simple bark-covered structure built within the original city-fort in 1642. However, what we were looking at was of one the first cathedrals in Canada resulting from the Gothic Revival style, designed by the Irish-American architect James O'Donnell and inaugurated in 1829.* The occasion was the annual meeting of the ACA that took place in this Canadian city in August of 1995. There were three of us in the group: Cynthia Stauffacher, John Badger, and yours truly. After our visit John uttered with some hidden regret in his voice, "We don't build these things any more," referring to the unique cathedral in front of us.

Although I did not verbally respond right away, the thought occurred to me immediately, "No, we build synchrotrons instead."

---

* I would like to thank my friend in Montreal, Ms. Laura Labrosse, for helping me track down the exact place of my recollections and for sending me the information about the *Notre Dame* Basilica.

The conversation went on to other things. Later in the day, I outlined in my mind the logic, by which, as a scientist, I thought human kind in general and individual societies in particular are better off devoting their resources to constructing and developing large experimental facilities such as synchrotrons. In a sentence, the argument typically goes as follows: the scientific findings and technological developments obtained exploiting those large experimental facilities would revert back into society, resulting in a better standard of living. J. D. Bernal, the influential crystallographer and scientist who pioneered the early study of protein crystals by X-ray diffraction, was a great believer in the role that science should play in the improvement of the human condition.[1] The societal commitment that is necessary to design and build dedicated storage rings—synchrotrons— for the production and use of X-rays for scientific investigation is enormous. The impact that they will have in our knowledge of the physical world in the next century and beyond is difficult to predict. As scientists, we do not question its critical role for the future of science, the public expense being analogous to an astronomical observatory, a particle accelerator or the space shuttle.

Those technological wonders are human artifacts that have practical value within the context of the secular and agnostic societies of today. Dedicated physicists, engineers, and instrument makers in accord with the scientific and technological goals of our societies build them; they are large projects funded by governments using tax revenues through their different agencies (**Plate 18.1**, lower left: Advance Photon Source, APS; upper right: Europeran Synchrotron Radiation Facility, ESRF).

One might argue that cathedrals, on the other hand, were built in the Middle Ages by towns or *"burgs"* strongly influenced by the religious beliefs and values dominant at the time. They constructed cathedrals as a way of *"rapprochement"* with the Lord Almighty that provided the compass of their lives. They were erected in the middle of town by masons, artisans, carpenters, and blacksmiths as a community effort, under the auspices of a bishop or religious leader. In our materialistic world of today we do not build those things any more. Or, do we?

One ordinary day, I received an invitation from Prof. H. Bernhart to the 5th European Congress of Medical Mycology* to present a lecture on the structure of a protein-cutting enzyme secreted by the pathogenic fungus *Candida albicans*. This organism can cause very severe infections in people with a weakened immune system, such as people undergoing chemotherapy or patients diagnosed as HIV positive. The congress was going to take place in the city of Dresden, Germany, in early spring. What caught my attention in the cover of the flier was an aerial view of what appeared to be a massive scaffold around a part of town that was basically in ruins. Who on earth would like to advertise a conference in a city that is undergoing massive reconstruction? What public relations firm had advised the organizers of the meeting to choose such a photo for the logo of the conference? I asked for the necessary permissions and responded that I would be delighted to participate. I filed the brochure and the ensuing correspondence in a manila folder and went on to other things.

The confirmation letters arrived a few months later. I submitted a title for my contribution and a second circular arrived with more details about the congress. This time a different view of the same scaffold around the walls of a building being restored was displayed on the cover of the brochure for the congress. Winter gave way to spring and in preparation for the meeting I went to the *Encyclopaedia Britannica* to read some more background on Dresden, the capital of Saxony in Germany. It was well recognized that for centuries the city had been a center for music and musicians on a European scale. Important episodes of the life of Bach, Handel, and Teleman had taken place in Dresden. The city had often resonated in the past with the operas by Carl Maria von Weber, Richard Wagner, and Richard Strauss. In addition to the unique examples of German baroque and rococo architecture, the picture galleries in the city included many works by old mas-

---

* I am grateful to Prof. Dr. Hannelore Bernhardt for her invitation to the 5th ECMM and for her hospitality during my visit to Dresden. I am also grateful for the enlarged view of the cover of the program for the Congress.

ters such as Holbein, Cranach, Vermeer, Rembrandt, Hals, van Dyck, Rubens, Botticelli, and Canaletto. All of these were the "treasures of Dresden."[2]

Unfortunately, I also read the tragic details of the February 13–14, 1945, bombing that reduced the "Florence on the Elbe" to rubble and ruins.[2] The American writer Kurt Vonnegut Jr. survived the firebombing of the city as a prisoner of war with another infantryman Henry J. Leclair in an underground meat locker, within the city slaughterhouse fifty feet underground. His chronicle of this infamous episode of World War II is iconoclastic, irreverent, and cynic, but at the same time is full of a profound human empathy.[3] Within this historical framework, I began to understand the brochures of the conference a little bit better.

However, it was only when I set my footsteps in the city; when I walked down its streets and avenues and strolled down the *"Altstadt"*; when I listened to music in the fully restored Semper Opera; when I indeed saw the raising walls of the *"Frauenkirche"* surrounded by a scaffold, and compared the view with old photographs of the city—only then did I fully comprehended the significance of those proud walls growing inch by inch towards the full *wiederaufbau* of the city symbol from the devastation of World War II. Then, I fully understood why the photographs of the restoration were being proudly displayed in the posters announcing the Congress.

A few months after I wrote these ruminations regarding the restoration of the "Florence on the Elbe" by their proud citizens—by then almost the end of twentieth century—a ray of hope appeared in the horizon. Prof. Günter Blobel (Rockefeller University, NY), winner of the 1999 Nobel Prize in Medicine and Physiology, announced that he would donate the entire monetary award for the restoration of the *Frauenkirche* and a synagogue in Dresden.[4] According to the report, Dr. Blobel watched the devastation of the city as a child of eight from a few miles away and is the founder of a U.S. organization named "Friends of Dresden" devoted to the restoration of the artistic treasures of that German city (**Plate 18.1**, lower right).

Upon further reflection, it also came to my mind the efforts to complete the cathedral, or "Templo de la Sagrada Familia" ("Temple of the Sacred Family"), in Barcelona, Spain, the magnificent design of the Catalan architect Antoni Gaudí i Cornet (1852–1926). This cathedral, more than any other that I know, represents an example of a basilica conceived, designed and being constructed in our time. Gaudí was not the original architect when the first stone was laid in 1882. However, he accepted the commission in 1883 and would devote his entire life to it. The original architectural design of the temple is unique in many ways but this is not the place to expound on the details of the plans. I would like to invite the reader to visit it. Nonetheless, I want to focus on three aspects of this masterpiece.[*]

First, the abundance of organic and biological forms in the design.[5] Antoni Gaudí was the foremost representative of a school of architectural design (Art Nouveau in France or Modernismo in Catalonia) that reacted against the generic, geometric, and linear forms of architecture at the beginnings of the 20th century. As to the overall design of the Sagrada Familia, Gaudí said that the nave, aisles, and the vaults of the basilica should be "like a forest" with light entering and flooding the windows. A forest accentuated by the twelve resonating bell towers reaching over 300 feet from the floor, four on each of the main facades and representing the apostles. Natural forms are displayed throughout; they are part of the crypt, the gargoyles, and spires of the apse. Plants and animals in the Nativity facade and fruits in the windows represent the different seasons of the year. A well-known detail is the presence of a turtle at the base of a column within the Portal of the Nativity façade. The poor creature seems to support the entire structure. Second, color is ever present in Gaudí's architecture and the Sagrada Familia is no exception The stained glass windows inside the temple and especially the mosaics and tiles as the terminations of the bell towers are fingerprints of the work of Gaudí. Finally we have the parish schools that were built inside the

---

[*] The information is from brochures and publications about the "Temple of the Sacred Family" sent to me by my dear friends Dr. Jaume Olivella and Dr. María Angels García-Bach from Barcelona.

cathedral in 1909. As the architect expressed it: "by the side of the Church, the people will receive education and culture." This last detail makes the cathedral truly a community building. Although the schools were destroyed and part of the church sacked during the Spanish Civil War (1936–1939), the continuation and eventual completion of this masterpiece by the 21st century are on track. The financing is drawn from private contributions from all over the world and the work is about half completed. I can only wish that when the completed work is dedicated in the current century the name is changed from "Temple of the Sacred Family" to "Temple of the Human Family" (*Templo de la Familia Humana*). This minor change will put this magnificent cathedral in its rightful place as an icon of "*rapprochement*" of all human beings irrespective of their individual faiths and credos (**Plate 18.1**, upper left).

I should close these lines mentioning a scientific event related to synchrotrons that will have social, political and possibly historical implications. When time came to decommission the seventeen-year-old German synchrotron BESSY I (*Berliner Electronen-Speicherring für Synchrotronstralung*), Herman Winick from the Stanford Linear Accelerator Center suggested the possibility of donating it to the Middle East region to foster scientific collaboration, and possibly as a seed for enduring peace. The idea has taken hold and UNESCO has given it a strong support. The plan would be to build an upgraded version and make it a centerpiece of collaborative research in various fields, among scientists from traditionally unfriendly nations.[6] The eleven country members of SESAME (Synchrotron-light for Experimental Science and Applications in the Middle East) agreed on Christmas Eve 1999 to contribute the necessary funds to get the project started,[7] and Jordan had been tentatively selected as a first-choice site for the facility.[8] A location in Yerevan, Armenia (former Soviet Union), was chosen as a second choice. Besides technological artifacts, synchrotrons might prove to be instruments for peace in years to come.[9] It is not first time that rings have been used as symbols of peace and cooperation: remember the Olympic games.

After putting all these recollections together, I realized that our casual, off-side and almost snobbish remark of a few years ago was simplistic and only the reflection of one facet of the human persona. Our brain and our curiosity, combined with our internal logic, need to devise and construct probes to study the material world, in a constant struggle to understand its internal machinery and improve the human condition. Inextricably linked with this part of our minds—what we call our hearts and spirit—is the need to imagine, dream, write, play, paint, sculpt, chisel, mold, and bend to overcome our fragility; to transcend our ephemeral nature; to make sense of our temporary passage through this earth. Then, let it be both: synchrotrons and cathedrals for a friendlier, more generous, increasingly fraternal, and peaceful 21st Century, for all the inhabitants across planet Earth.

In closing, after the tragic and horrific events of September 11, 2001, in New York City, it might be appropriate to bring back an obscure quote from the great humanist and scientist Louis Pasteur (1822–1895) who, without having known the two World Wars and many other abominable acts against humanity, wrote in 1878:[*]

> "I beseech you to take interest in these sacred domains so expressively called laboratories. Ask that there be more and that they be adorned for these are the temples of the future, wealth and well-being. It is here that humanity will grow, strengthen and improve. Here, humanity will learn to read progress and individual harmony in the works of nature, while humanity's own works are all too often those of barbarism, fanaticism and destruction."

---

[*] I thank my personal friend Ruth Gyure for bringing this quote to my attention in the aftermath of the tragedy of New York City, September 11, 2001. The English translation from the Pasteur Foundation in New York City, (www.pasteurfoundation.org). The original French is from reference.[11] Reproduced here with permission from the Pasteur Institute. The assistance of Ms. Mirdza E. Berzins from the Lake Forest Library is appreciated in tracking down the source of this reference.

VII

practical applications

19

Molecular Docking and the Broken Heart

P aul Ehrlich (1854–1915), widely recognized as the founder of modern medicinal chemistry, coined the word "chemotherapy" in 1891 to refer to the curative effect of certain man-made chemical entities. In his view, critical to the success in synthesizing these compounds were the three Gs: *Geduld*, *Geld*, *Glück* (patience, money, and luck). After 606 exhausting trials, Ehrlich developed in 1910 an arsenic compound ("salvarsan"), which allowed the effective treatment of syphilis.[1] This was the first man-made compound that cured an infectious disease in man. Since then the pharmaceutical industry has continued to discover, synthesize, test, develop, and market novel "drugs" with a tremendous impact on our societal and individual well-being.

The Food and Drug Administration (FDA) was created in the U.S. in 1938 to enforce the Federal Food, Drug, and Cosmetic Act, which put tighter restraints on industry practices. The international tragedies resulting from the use of thalidomide in Europe led to the 1962 amendment of the legislation (Kefauver-Humphrey Drug Amendment signed by J. F. Kennedy) and caused drastic restrictions in clinical investigations and stringent requirements on any experiment involving human subjects.[2] Consequently, the finding of truly novel and safe pharmaceutical entities slowed down considerably. Besides, the cost of putting a new and effective drug on the market had risen to figures between $200 and $300 million, with an elapsed time ranging from 10 to 12 years of devoted research effort distributed among discovery, development, and clinical testing. In an atmosphere of diminishing returns, the pharmaceutical industry found itself running low in the three Gs. At the same time, during the 1950s and 1960s, there was a dramatic increase in our knowledge of the enzymatic reactions occurring in living organisms. This pool of biochemical information added a new dimension to the understanding of the relations between the chemical reactions necessary to sustain life, the enzyme catalysts participating in these reactions, and the small chemical entities that participated and altered these chemical processes.

It is a tribute to the internal dynamic and ingenuity of the American and other western societies, that the crystallographer's technical—and some might say esoteric—expertise soon found a niche inside the walls of corporate America to aid in the process of inventing and designing pharmaceutical drugs. The premise is self-evident: protein molecules through their enzymatic or regulatory activity control most biological processes. If you can inhibit those enzymatic processes, or if you can modulate their actions, then you can intervene with biological processes and probably will achieve therapeutic results.

During the early 1980s a target of choice was renin, an aspartic proteinase, which is part of the series of reactions regulating blood pressure in humans. A myriad of compounds were synthesized to block the activity of renin and several of those compounds were crystallized and analyzed as complexes with renin or with the archetypal aspartic proteinase, pepsin. In spite of the difficulties of obtaining good quality renin samples to perform the crystallographic studies, the renin target validated the structural approach and the concept of "structure-aided drug design" was a considerable driving force behind the practical applications of protein crystallography.

Strikingly, the epidemic that emerged in the late 80s featured a malignant virus (HIV) that needed another aspartic protease, known as HIV protease, to complete its life cycle. This molecular target was a much smaller dimeric protein that could be produced in large quantities by DNA recombinant methods (see Chapter 21, *"The Soft Engineers"*).

It can be crystallized either by itself or in complex with a multitude of "designed compounds" and dedicated efforts resulted in well-diffracting crystals in laboratories all over the world. The structures of over 200 HIV-protease:inhibitor complexes have been analyzed world-wide. I am proud to say that several years ago the compound that turned out to be ritonavir (NORVIR™) went through our laboratory and my colleague Dr. Chang Park provided valuable crystallographic data (PDB entry 1HXW) to the project that designed this successful drug against the AIDS epidemic.[3] Initially, a series of inhibitors was designed based on the dimeric structure of the HIV protease.[4] This breakthrough was followed by extensive optimization of the activity, physico-chemical properties, metabolic and pharmacological behavior and eventually led to the identification of ritonavir. The introduction of HIV protease inhibitors has resulted in the development of second-generation potent-combination protocols that are effective in reducing the amount of virus in the blood of AIDS patients to undetectable levels[5,6] (Plate 19.1).

In spite of this dramatic success, drug design is a very complex endeavor. The parameters and variables that need to be optimized are related to the molecular interactions between the "drug" and its "target"; its toxicity, oral bioavailability, and also to its pharmacokinetics (i.e., the metabolic degradation or average life time of the active compound in the blood of the patient). Well before the three-dimensional structure of the target enzymes was available, quantitative concepts had been defined to address these properties at the molecular level, and approaches to correlate these molecular properties with their *in vivo* activity were referred to as Quantitative Structure Activity Relationships (QSAR). Figures for critical parameters such as inhibition constants in the appropriate *in vitro* and *in vivo* assays (K_i, IC_{50}, MIC_{90}, EC_{50}), partition coefficients between *n*-butanol or *n*-octanol and water (LogP, CLogP), and others fill innumerable boxes of immense spreadsheets in the minds and personal computers of scientists involved in drug discovery. These quantities are the beacons necessary to successfully navigate the treacherous waters of any drug-design effort. Crystal structures of the target macromolecules and of their complexes with active molecules provide a very important piece of the puzzle but not the only one. Although drug design is a team effort *par excellence* requiring the efforts and expertise of many investigators, compounds are synthesized by medicinal chemists and assayed by biochemists and biologists; both efforts are of pivotal importance to the final outcome.

Even knowing the three-dimensional structure of receptor:inhibitor complexes, theoretical calculations are still inadequate to obtain reliable numbers for some (e.g., K_i) of the macroscopic quantities defined by the above concepts. Indeed, others (e.g., MIC_{90}) cannot be calculated within the structural framework. Amazingly, the simplest component of all, water, is the most difficult to incorporate successfully in our calculations. Available methods are currently unable to tackle at the molecular level issues such as absorption by the different tissue barriers, bioavailability, and pharmacokinetics. Super inhibitors (pico-molar or better), that we can discern in exquisite detail in the protein:inhibitor crystals, fail tomorrow as drug candidates because of permeability and accessibility problems or because of various associated toxicities.

Further challenges lie ahead with the advent of the gargantuan amounts of information provided by the genomics revolution. We are going to find more and more often that strings of characters *in silico* do not reveal themselves as proteins with an obvious activity *in vivo*. Many of the ones, for which an enzymatic activity can be assigned, will turn out to have subtle and pleiotropic effects in tissues and organs, complicating the issues of target validation, definition, selectivity, and suitability. Still others will be proteins associated with or bound to cell membranes, which will make them very difficult to manipulate separated from their native milieu. These unknowns will drive sophisticated biological, genetic, and biochemical experiments well into the 21st century.

In addition, the "hyper-rational" dream of designing drugs and predicting their activities *ab initio* will attempt to exploit the sheer computational power of computers such as the 32-node supercomputer (IBM RS/6000*SP) that executed "Deep Blue", the landmark computer program that crushed Garry Kasparov in the momentous man-machine chess tournament a few years ago. Unfortunately, the rules that govern the energy (ΔG), enthalpy (ΔH), and entropy ($T\Delta S$) terms of the interactions between the molecules of our dreams and their putative targets are not as well-defined as the chess moves. In spite of our ingenious genetic algorithms and our docking searches, the "value functions" are still difficult to quantify and to translate into positive gains.

I cannot resist finishing these lines about structure-aided drug design with a quotation from the novel *Written on the Body* by the British writer Jeanette Winterson (b. 1959).

"Molecular docking is a serious challenge for bio-chemists. There are many ways to fit molecules together but only a few juxtapositions that bring them close enough to bond. On a molecular level success may mean discovering what synthetic structure, what chemical, will form a union with, say, the protein shape on a tumor cell. If you make this high-risk jigsaw work you may have found a cure for carcinoma. But molecules and the human beings they are part of exist in a universe of possibility. We touch one another, bond and break, drift away on force-fields we don't understand"[...]*

The soliloquy comes from the brain of the main character in the novel, as she** debates whether staying close to her lover will heal her broken heart or could, in fact, result in a catastrophic and ruinous experiment. I will close these reflections in a similar vein.

We, humans, have made tremendous scientific and technological progress. Some members of our species have walked on the moon,

---

* I owe this quotation to my wife. She discovered it in the book by Jeanette Winterson.[7] Reprinted by permission of International Creative Management, Inc. Copyright © 1994 by Jeanette Winterson. First published by Vin.

** When the original version of this essay was published in the *PDB Newsletter* (April 1998), Dr. Bernhard Rupp complained that in no place did the author of the novel make a statement regarding the gender of the main character. This is correct. I responded appropriately to this comment (see July 1998 *Newsletter*). My response was that it is easier literally to write the sentence using one gender. At the same time, since I never declared the name of the lover, the quote might be taken as being sexless from the other side.

cloned mammals, cleaved an atomic nucleus, and discovered new planets, galaxies, black holes, and other astronomical objects. After 300 years, even Fermat's last theorem has been proven by an inquisitive and determined member of our species. Others have mapped the structure of the common cold virus in atomic detail. Aided by the knowledge of the biological machinery of the AIDS virus at the atomic level, we have been able to design effective drugs to fit molecules of this virulent pathogen, and we are now treating what were considered untreatable diseases only a few years ago. Our data base of structural knowledge has expanded enormously and will continue to do so allowing us to make new strides in the constant struggle against aging, disease, pain, and deformity.

However, the force fields that govern our interpersonal relationships are beyond our control. The gradients controlling our passions, our desires, and our affections defy our comprehension and are not amenable to analytical tools. Far from our rational understanding are the tribal winds that carry hatred, prejudice, bigotry, injustice, and war. The human hearts expand and shrink, thrive or suffer subject to force fields very different from the ones controlling the molecular docking of an inhibitor to its target or a substrate to its receptor. I wish the magic of crystals could help alleviate some of these problems.

20

Itsy, Bitsy Spider

"What's so miraculous about a spider's web?" said Mrs. Arable.
"I don't see why you say a web is a miracle—it's just a web."
"Ever try to spin one?" asked Dr. Dorian.
Charlotte's Web by E. B. White

t dawned a morning like any other one, with nothing ahead of us but hard work. I was part of a crew of migrant workers in the vineyards of *St. Etienne de Lisse*, approximately eighty kilometers east of Bordeaux, France. The fields in this small village and its immediate surroundings produced the grapes that experienced artisans have learned to transform into the superb red wines of the *appellation d'origine "St. Emilion"*. The name originates from a French town only a few kilometers away, famous for its wines since medieval times. During two weeks in mid-September, I used to take a train from my hometown in the northern part of the Castillian plateau and join groups of temporary laborers in the vineyards of France to make some money for college. The work was backbreaking but it gave me the opportunity to learn French, enjoy the countryside, and, occasionally in the evenings, go to the local fairs and admire the local beauties. At the end of the harvest, as a bonus to the salary, we would get a couple of bottles of wine from the previous vintage. In later years, I saved those bottles for my future father-in-law D. Juan Manterola.

Although it began plain and unpretentious, something happened that morning that I have always remembered and that marked my enormous admiration for the inner-works of biological structure. It was cool and humid. I finished harvesting the rich and abundant grapes on my side of a long row of vines. Then, as I stood up to relieve my back, I saw the largest and most beautiful spider web that I had ever seen in my life: it was a magnificent specimen of the orb-web type. The dew had condensed in tiny droplets regularly attached to the delicate curves, like iridescent pearls of various sizes, and the sun, shining in the background, created a splendid view that embraced a kaleidoscope of interconnecting minute rainbows. I rushed to our *patron* and I asked him permission to go to the barracks and bring my old camera to take a picture. The only black and white print that I had of that fragile structure, and of that magic moment, I gave as a present to my old girlfriend, who apparently destroyed it by fire when she married somebody else. So it goes!

Nevertheless, the image was vividly stored in some obscure recess of my brain. Aren't we all fascinated by the exquisite design of biological structures? From the humble and fragile dandelion seeds to the robust acorns; from the delicate spider webs to the giant sequoia trees; from the minuscule external sculptures of the diatoms that drift in the sea, to the gargantuan skeletons of the extinct dinosaurs; the structures created by organisms surpass in beauty and efficiency anything created by the human hand. How is this possible? Do organisms know special rules of architecture and design that escape the sophistication of the human brain? Do the atoms that constitute the living matter know any special kind of biological geometry that allows them to create those intricate designs? Yes they do, and yes they don't.

The predominant atomic components of the living world are rather different from the atomic ingredients of human technology. Organisms create their unique design using predominantly atoms of carbon, nitrogen, oxygen, hydrogen, sulfur, and phosphorus with minor amounts of inorganic ions such as Ca^{++}, Na^+, K^+, Mg^{++}, and Zn^{++}. The hard metals such as iron (or steel) and copper, so common in human technology, do not play a major role in biological structures. In very small amounts, they are very important for metabolic processes but the exploitation of metals and their alloys (i.e., bronze) to support stress is exclusively a product of human technology, especially after the industrial revolution. In fact, J. E. Gordon, an authority in materials science, has suggested that the industrial revolution was in some respects too successful because it swept away as inadequate and old-fashioned the use of other traditional materials and manufacturing processes:

> "Before the Industrial Revolution, technology was based on a wide range of materials: on various kinds of wood, stone, and leather and on rope and cloth made from various natural fibers. Metals such as iron and copper were used too, but only to a moderate extent. The craftsmen who fabricated these materials often had little formal education, and they worked and thought in traditional 'unscientific' ways. However, the nature of their trades forced upon them a certain humility toward the physical substances they sought to master.
>
> During the nineteenth century much of this traditional technological culture was swept away, and most of what remained was disparaged as out of date. [...] There would be no need to endure the emotional discomforts of fundamental and drastic rethinking; certainly there would be no call to reevaluate the work of the old-fashioned shipwrights and coachbuilders."[1]

This new technology should not be dismissed as not having its own purpose, lightness, and beauty. Examples are the elegant bridge over the Douro River at Oporto (Portugal, 1877) and the unique landmark in the city of Paris (1889), both the achievement of the French engineer Alexandre-Gustave Eiffel (1832–1923). Nonetheless, a sense of knowing-it-all became prevalent among "iron engineers" who drew little inspiration from the fascinating structures of medicine and biology, or demonstrated very little interest in the natural solution to complex engineering problems.

The need for better insulating materials in the young and dynamic electrical industry forced the search for new materials. The introduction of plastics (i.e., bakelite and fiberglass) and organic polymers such as nylon in the first half of the 20th century brought human-made materials one-step closer to the elegance of biological materials. Expanding knowledge of chemistry was beginning to make it possible to design materials based upon their atomic components. In other words, it became possible to synthesize materials, taking into consideration their atomic structure and building from the inside towards the outside, as opposed to modifying a pre-existing material (i.e., from the outside towards the inside).

Basically, humans have transformed and created their environment by cutting, casting, molding, and bending available materials in certain shapes and forms. Of particular importance in our lives are the materials that we use to construct our homes, offices and public buildings, and the artifacts that make our lives more comfortable. In addition, through history books and archaeological excavations we have also learned about the materials that humankind used to make war (i.e., stone, bronze, and iron) and to conquer and destroy one another. The role, which softer materials such as leather, obtained from animal hides, have played in human history has only been briefly outlined by the most insightful and astute observers of the development of human societies.[2] I wonder how much of the world would Alexander the Great have conquered without leather sandals and saddles, or in the absence of leather harnesses for the horses in the chariots of his troops. The advent of organic polymers as a result of the growth of chemical synthesis and manufacturing created new materials but the resulting artifacts were still produced by imposing a form on a pre-existing mass or bulk of material.

Interest in the chemical composition and function of the polymers that supported and maintained life was a natural extension of the burgeoning biochemical knowledge of the first half of the twentieth century. The unraveling of the three-dimensional structure of DNA—the polymer carrying the genetic information—by J. D. Watson (b. 1928) and F. H. C. Crick (b. 1916), aided by the X-diffraction patterns from oriented DNA fibers obtained by the late Rosalind E. Franklin (1920–1958) and Maurice F. Wilkins (b. 1916) has been told countless times.[3-5] I want to ruminate on the study of other, less glamorous, biopolymers of interest using the methodology of diffraction of X-rays on partially crystalline (i.e., para-crystalline) materials such as fibers. I am referring to many of the proteins that play critical roles in building the structural materials that support life or that are produced by organisms in their multiple adaptations to the various environments. Among these biological materials are hair, silk, wool, horn, and fingernail, to name just a few of the more accessible ones. These more humble and certainly less "selfish" chemical entities are also marvels of design and efficiency in their own right. For many years, they held the secret of one of the most intriguing puzzles of structural biochemistry in the first half of the twentieth century: What was the three-dimensional structure of proteins? How did the polypeptide chain fold in three-dimensions?

A technical issue needs to be presented to understand the difficulty of the problem. As opposed to single crystals, fibers, and other para-crystalline materials do not produce a rich diffraction pattern when exposed to intense X-rays. Their atomic regularities are only reflected in a few strong—but diffuse—spots when exposed to intense X-rays. Those few reflections appear predominantly along the two perpendicular directions of the diffraction and are referred to as *equatorial* (i.e., horizontal, through the origin) or *meridional* (i.e., vertical), typically associated with the distance in Ångstroms that produced that reflection. Having only a few dominant structural parameters, the construction of an acceptable model of the folding of the polypeptide chain had to rely on hypotheses about the underlying chemical nature of the protein chain. The late chemist Linus Pauling (1901–1994) was the undisputed master of this realm.

Diffraction patterns from keratin, the principal protein component of wool, hair, horn, and nail, had been published by William T. Astbury (1898–1961) in the early thirties. Educated at Cambridge, he was a member of the old school of crystallographers who had settled at the University of Leeds (county of Yorkshire) in 1928, within the heart of the northern English textile industry and had been trained by the elder Bragg. Although scarce, the diffraction data obtained by Astbury from these biological polymers provided the first insight into the structural constraints imposed on the structure of proteins.

Moreover, he was able to show that distinct changes in the structural parameters occurred between the resting and stretched wool and he referred them as α- and β-keratin. The diffraction patterns gave suggestive indications of a helical coiling pattern—much like an old-fashioned telephone cord, before the advent of mobile phones—with a distance of 5.1 Å between any two consecutive turns (i.e., helical pitch of 5.1 Å).

There is quite of bit of history and "folklore" on the unraveling of the structure of the α-helix, one of the basic folding patterns of proteins;[4] I'll provide a concise version. Robert B. Corey (1897–1971) joined Pauling's laboratory at the California Institute of Technology in 1937 and they began a series of structural studies on simple molecules made up of few amino acids, the building blocks of proteins. Careful analysis of the structure of these small peptides revealed the essentially invariant dimensions of the different bonds and angles holding the atoms of the amino acids together. In particular, they realized that the peptide bond—the bond connecting two amino acids (see **Glossary**)—had to be planar: all the four atoms making up the bond C=O, N-H will have to lie on the same plane. This pivotal observation, together with Pauling's chemical modeling wizardry, suggested that the polypeptide chain of proteins would indeed coil in a helical structure but the pitch should be 5.4 Å instead of 5.1. The new model aligned the double-bonded oxygen of one amino acid with the hydrogen attached to the nitrogen of the residue one full turn above it along the helical axis: C=O---H-N. This elegant solution to the folding of the polypeptide chain came to be recognized as the α-helix: an elegant tube of atoms (6.8 Å in diameter) embedded on a cylindrical surface, held together by the sharing of hydrogen atoms among the constituent amino acids. Whimsically, the number of residues per turn of the helix was non-integral, a fact that had eluded many of the best minds in the field; it turned out to be 3.6 residues per turn (**Plate 20.1**).

The film *The Man in the White Suit* (1951, directed by A. MacKendrick and based on a play with the same title by Rogert MacDougall) is a unique science fiction film from the 1950s dealing with the repercussions of the discovery of a special synthetic fiber, which was eternally durable. An eccentric British textile chemist, played by a very young Sir Alec Guinness, invents a fiber using amino acids with revolutionary properties: it never stains and will never wear out. As amazing as the premise might appear, it is no worse that the claims made by many other science-fiction films that have followed in later years. The value and moral of the story derives from the social interweaving of the scientist, capital, and labor in the face of such a stunning invention. In the end, the invention died out because of economics. This film is one of many films that my father "forced me" to see during my adolescent years, to learn to appreciate good filmmaking. It is a classical movie that I did not fully appreciate at the time. However, I recently rented a videotape (black and white, of course) and enjoyed it tremendously; there are several fascinating scenes and situations that I could not possibly recount here. I should point out however that the fiber was so strong that in one of the scenes the protagonist is able to escape down a window using the fiber as his only means of support.

In retrospect, I wonder whether the original play was partly inspired by work of the British textile industry and in particular the Courtaulds company. William Astbury collaborated with them, and in the 1940s and early 1950s this company synthesized a fibrous polypeptide that was somewhat different in composition to keratin and whose diffraction pattern did not immediately exhibit the characteristics of an α-helix. When the first diffraction data from these artificial fibers were published, the trained eye of Linus Pauling recognized immediately that the fibers were α-helix packed together like a handful of cables in an hexagonal array. The data also provided the final nail in the coffin of Astbury's helical model because the 5.1 Å peak was missing.[6]

I do not know scientifically or technically what happened to this new synthetic fiber synthesized and characterized by the Courtaulds company. One can dream that as in the story of the "man in the white suit", capital and labor attempted to terminate the manufacture of the material because such a revolutionary fiber would ruin the industry.

Perhaps more realistically, technical issues that appeared later prevented the realization of the dream of Sidney Stratton, the protagonist in the story; I truly would like to know. In either case, in the film Alec Guinness wore an impeccable (literally without sin or stain) white garment, more similar to silk than to wool. The atomic structure of these two natural fibers is very different, although both are protein-based.

In their imaginative papers, Pauling and Corey proposed also another basic structure for the polypeptide chain where the chain was almost fully extended, forming a wavy sheet with shallow peaks and valleys. In this structure two or more chains run in opposite directions (anti-parallel) to each other and the "hydrogen bridges" between the C=O and the H-N atoms are perpendicular to the chain direction, forming between two separate chains or strands. They called it an anti-parallel pleated sheet, but it is normally referred to as a β-sheet. The resulting fiber is very strong, practically impossible to stretch, and is the main structural feature of the different silk from spiders and the silkworm *Bombyx mori* (**Plate 20.2a,b**).

α-helices and layers of β-sheet constitute the bone and fabric of protein architecture. Within morphological proteins, they are used in larger stretches combined in complex ways to suit the properties of the material. All types of wool and hair are abundant in keratin, which is essentially built of α-helical rods as basic units. Feathers in birds are intricate designs of β-sheet. The different varieties of silk are the best examples of β-sheet and collagen, the predominant component of connective tissue, is a coiled coil of three α-helices. Proteins that do not play a structural role in organisms are normally referred to as globular because they can be isolated as discrete entities and have a more compact shape. Within this class of proteins, Nature has learned to combine the two basic structural elements in a myriad of architectural themes and variations to produce unique chemical entities that can: catalyze reactions at amazing speeds; transport oxygen by holding it delicately in an flexible, atomic basket (i.e., hemoglobins); generate chemical and mechanical energy (muscle proteins); trap sunlight and channel it to direct chemical synthesis (photosynthesis and antennae proteins); produce mechanical energy without thermal differences (cell motions); receive and transmit chemical and ionic signals (receptors and neuronal signals), and many more. As we know it on this earth, proteins make life possible.

We are all very fond of telling our children the fable of the *Three Little Pigs* to convey upon them the importance of a solid dwelling to protect ourselves from our real or imaginary enemies. We do recognize the beauty and harmony of a well constructed home or building (i.e., a cathedral). Nevertheless, because of time or economic pressures we compromise, we cut corners and mass produce our homes in America with the prosaic wooden "two-by-fours," and enclose the space with compressed boards of particulate wood. Personally, I have always dreamed of living in a house constructed of nacre, the material inside of mother of pearl and shells: large, elegantly curved plates, solid and harmonious on the outside, but soft and delicate in the inside. As a bonus, it will not require any maintenance, will last for a long time and if we design it right, the shell may expand as our family grows. Such a material is still far into the future and may be prove to be unsuitable for the construction industry. Yet, marine conchs have always been for me the most exquisite examples of materials produced by organisms. I never get tired of admiring their seemingly infinite variety of forms, feeling their various textures, and being dazzled by the kaleidoscope of colors and patterns.

Unlike materials designed by humans, marine shells combine the soft textures of organic materials (using C, O, N, and S) with the strength of minerals and rocks containing limestone, calcium carbonates, phosphates, etc.). Their secret, which we have not yet mastered, combines layering an organic matrix—probably of protein nature—as a substratum to guide the deposition in a pre-ordained manner of the hard ions such as Ca^{++}, phosphate, SiO_2, and even ferrous (Fe^{++}) or ferric (Fe^{3+}) iron. This is much like bone, which starts life in the embryo as the tough but flexible protein called collagen, and is later covered by rigid fibers of the inorganic calcium compound hydroxyapatite ($Ca_3(PO_4)_3OH$). One recent report[7] documents precisely this phenomenon in the formation of the siliceous skeleton of diatoms.

A second paper[8] describes the structural basis for the unique fracture properties of the shell of *Strombus gigas*, the giant pink Queen conch from Caribbean habitats. This heavily mineralized material (~ 99% aragonite, a form of calcium carbonate) contains only about 1% of its constitutive matter as an organic matrix distributed amongst several inorganic layers. This arrangement results in a composite material with a fracture toughness 100–1,000 fold superior to the pure mineral alone.

I am certain that in the years to come the systematic molecular study of the materials created by even the most humble of organisms —either plant or animal, the distinction is irrelevant—will teach us many valuable lessons.[9,10] Insights that will help us on how to best be part of our natural environment, in a soft, gentle and accommodating manner, as opposed to the harsh and dominant way that we have become so accustomed to in the past. I would like to end by paraphrasing a quote from Janine Benyus in her book *Biomimicry*[11,12] referring to silk: "a 380-million-year-old fiber with a twenty-first Century future." Biopolymers or materials produced by organisms are "400-million-year-old materials with an appeal to twenty-first Century technology". We have much to learn from free-floating microscopic diatoms; ferocious-looking but humble spiders; secluded, yet elegant chambered nautilus, and mighty but gentle oak trees.

21

The Soft Engineers: Protein Engineers

Traditional engineers and architects can modify our physical environment in visible and dramatic ways: bridges, express-ways, skyscrapers, and other massive structures. They often wear hard hats in their working environment and the materials of their trade are iron, steel, and concrete. The advances in our under-standing of the chemical and genetic make-up of organisms has resulted in the appearance of a new breed of engineers that I'll refer to as "soft engineers". Soft engineers wear white lab coats and use cells as microscopic factories. Their materials are biological poly-mers such as DNA, RNA, and a wide variety of enzymes to amplify, cut, digest, and re-link DNA molecules. All these reagents and reac-tants can be contained within small plastic test tubes, no larger than a few centimeters. The product of their labors is dramatically differ-ent from the massive structures built by the "hard engineers."

I'll present a concise summary of the concepts and *modus operan-di* of genetic engineers in three different ways, using three separate paragraphs. First, I'll use a literary metaphor, using the terminology of word-processing with the computer; this will be for the word-inclined person. Second, I'll make an attempt to use a more artistic analogy, using a tangible three-dimensional object and words used by people in the arts and crafts community. Finally, I'll go for the full-fledged technical jargon, using the appropriate generic names for the enzymes and reagents; definitions of the terms can be found in the **Glossary**. The explanation will be progressively more complex as we go from the first to the last paragraph. You can read through all the three descriptions or skip the paragraphs until you find the one with which you feel more comfortable.

Think of the information contained in the genes of the *Escherichia coli* (*E. coli*) bacterium—its genome—as a long string of characters (i.e., letters and numbers), divided into words, sentences, and paragraphs. Those encoded words are translated by the cell machinery into proteins and other large molecules that the bacterium uses to perform its day-to-day functions: searching for food, absorbing and metabolizing nutrients, synthesizing its own constituent molecules, and reproducing. This string of characters can be extracted from the bacterium and be processed separately to change those messages or instructions. Selected characters can be changed either to correct mistakes or to slightly modify the instructions. Full words or sentences can be cut and replaced by different ones, or pasted somewhere in the text. Most importantly, entire new paragraphs can be inserted with a message encoding a protein that is normally not produced by the *E. coli* bacterium. For instance, we can insert at the appropriate place and subject to the correct controls, the sequence of words required to encode for the protein human insulin, of critical importance for diabetic therapy. This modified genome, can then be put back into the inside of a population of cells, which appropriately grown will produce relatively large (milligrams to grams) quantities of the desired protein.

A three-dimensional analogy occurred to me when our son's necklace broke. Imagine the circular genome of the bacterium as a colored necklace containing beads in two different threads, twisted around each other. Since four letters carry the information of the genes, consider four different colors of beads: Aqua, Turquoise, Green, and Cobalt-blue (A, T, G, and C). Although the sequence of colors along each thread for the necklace is free to vary, the corresponding colors on the other strand is such that they form pairs: Aqua always goes with Turquoise (A-T or *vice versa*) and Green is necessarily matched by Cobalt-blue (G-C or the reverse). Purposely, I have chosen colors with first letters corresponding to the initials of the four chemical entities that one can find in all textbooks: Adenine, Thymine, Guanine, and Cytosine (see **Glossary**). This physical model makes it easier to understand why, when one is splicing fragments with a new sequence of colors into the circle, it is better to have rugged ends, whereby one strand is longer than the other and the missing beads can be restored by the color complementarity. Otherwise, it will be difficult to stick the piece end-to-end; just try it. With this physical analogy, the work of genetic engineers consists of replacing one or a few colored beads with others of different color; substitute certain fragments by others of similar length, or splice-in entire new sections in the necklace, making it a wider circle.

The genome of *E. coli* is circular and consists of a long double-stranded molecule of DNA, and can also sometimes contain smaller circular fragments called plasmids. The nucleotide sequence along each of the strands of the DNA molecule contains four chemical bases named Adenine (A), Thymine (T), Guanine (G), and Cytosine (C) in an unrestrained order. However, the complementarity between the bases requires that on the opposite strand A matches T (or *vice versa*) and G is met by C (or reverse). This complementarity permits a faithful replication of the genetic material, as was wittily noticed by Watson and Crick in their original paper. In addition, it ensures that the overall diameter of the cylinder encompassing the DNA fiber is essentially the same. Genetic engineers use enzymes called restriction enzymes to cut DNA molecules at specific places. Different restriction enzymes recognize unique sequences of DNA and can break DNA molecules in a variety of ways. There is a large variety of restriction enzymes, named using the first letters of the scientific name of the organism from which they were obtained. For instance, *Bam*HI (isolated from *Bacillus amyloliquefaciens* H). Among them *Hind*II (*Haemophilus influenzae* R_d) cuts the DNA resulting in blunt ends, while *Eco*RI (*Escherichia coli* RY13) leaves jagged (or sticky) ends. DNA ligases are enzymes capable of re-linking DNA fragments with either complementary ("sticky") or any type of blunt ends. Using these two classes of enzymes, molecular geneticists are able to cut and paste a gene of interest into plasmid DNA. These plasmids can then serve as a vehicle for introducing the altered genetic material into bacteria (**Figure 21.1**).

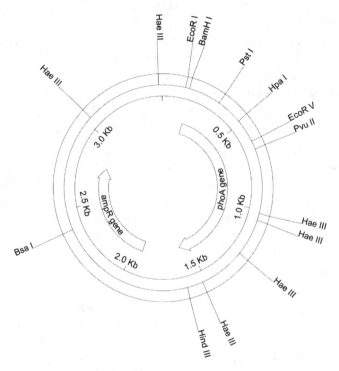

Figure 21.1. Schematic representation the genetic map of the *E. coli* plasmid used to produce the mutants of the enzyme alkaline phosphatase (phoA gene) with markers indicating the position of the different sites where various restriction enzymes can cleave and also the location of the ampR (ampicillin resistance) gene used as part of the plasmid. (Courtesy of Cathy Brennan, Abbott Laboratories).

Bacteria containing the re-engineered plasmid DNA are able to produce novel proteins. By suitable growth under controlled conditions, the modified bacteria can produce sizable quantities (milligrams to grams) of the desired proteins. It should be emphasized that the use of microorganisms to produce valuable chemical entities is not *per se* new. Special strains of the fungus *Penicillum* were used in the 40s to mass-produce the wonder antibiotic penicillin. The fermentation technology is commonly used in the pharmaceutical industry to manufacture large quantities of antibiotics, drugs or their important precursors. Novel is the use of bacteria with an **altered genetic material** to drive the process.

This technology is used in many laboratories to produce larger quantities of proteins that occur normally in minute quantities in the normal living cells, such as receptors, hormones, etc. This routine availability has made possible the explosion of novel structures of rare but extremely interesting proteins. Moreover, by the techniques of molecular engineering it is possible to replace individual amino acid residues (site-directed mutagenesis) in proteins. Of particular interest to protein crystallographers is the possibility of obtaining proteins whereby the sulfur-containing amino acid methionine can be replaced by the corresponding residue containing selenium. This nowadays common procedure, in combination with the availability of synchrotron sources, has accelerated dramatically the pace of the determination of protein structures (see Chapter 9 *"The Combs of the Wind"*).

Genetic engineering in combination with protein crystallography has been used to design proteins with novel properties of interest to the biomedical community. A case in point is the work that Guy Dodson, a disciple of Dorothy Crowfoot Hodgkin, and his colleagues at York University, who designed a more effective molecule of insulin. Insulin is a critical hormone that acts in conjunction with glucagon to regulate blood sugar levels. Insulin is the warning signal if the level is high, while glucagon sets the alarm if it is too low. After the work of F. G. Banting (1891–1941) and C. H. Best (1899–1978) in Toronto in the 1920s, the insulin molecule has long been a critical agent in the treatment diabetes. Insulin is a small protein that contains two separate chains of 21 (A-chain) and 30 (B) amino acids. It was crystallized in 1935 by Dorothy Hodgkin soon after she moved to Oxford,[1] but the complete three-dimensional structure was not unraveled until approximately thirty years later. Some of the difficulties were due to the fact that such a small protein has the tendency to aggregate first in groups of two (dimers), and later in clusters of six (three dimers, form an hexamer). Even though the active form *in vivo* is the single molecule (monomer), hexamers are present in the crystalline state, and the ion Zn^{++} is important to keep the three dimers together. Normally, injected insulin is hexameric and therefore the glucose level in the serum declines slowly due to the time it takes for the hexamer to break down into its individual monomers. Analysis of the three-dimensional structure of the hexamers of insulin in the crystal gave important clues to Guy Dodson and his colleagues as to how to design a variant of human insulin with faster action in patients. The structure suggested that replacing one uncharged amino acid residue in the B-chain (B28, a proline) by a charged residue (aspartic acid, containing a negative charge) will disturb the surface of contact between the A and B chains enough to expedite the hexamer-monomer conversion. Importantly, this change left the physiological activity of insulin unchanged.[2]

Single-residue mutations are extremely valuable for probing and dissecting the rate-enhancing properties of enzymes and to design more effective and stable catalysts. The three-dimensional structure of proteins involved in catalytic processes typically reveal a constellation of a few (three to five) critical amino acids, well conserved among enzymes of the same family, that are essential for the enzymatic action. Protein crystallographers, biochemists, and molecular biologists design variants of those enzymes with one or a few of those essential residues replaced by other functionally different amino acids. Analysis of the enzymatic properties of these novel enzymes helps them to understand the role that each residue plays in the overall reaction pathway. Of particular interest is the fact that some of those alterations may result in enzymes with faster reaction rates than the original enzyme. I was fortunate enough to participate (with my colleagues Wlodeck Mandecki, Cathy Brennan, Liqing Chen, and

Chris Dealwis) in one such project involving the enzyme alkaline phosphatase. This combined effort resulted in a better enzyme to be used in commercial diagnostic assays.[3] The replacement of residue numbered 101 in the polypeptide chain (aspartic101, a negatively charged residue in the native enzyme) by serine (an uncharged residue written as Asp101→Ser or Asp101Ser) resulted in an enzyme with approximately 30-fold faster activity. In the process, we also discovered a variant of the enzyme that would loose the bound magnesium ions (Mg^{++}) very readily, altering the structure ever so slightly and causing the corresponding crystals to crack[4] (**Plates 5.1c-d** and **21.1**).

The study of crystals of the native and mutant Asp101→Ser enzymes suggested an atomic explanation for the novel properties of the modified enzyme (**Plate 21.1**). The replacement of aspartic101 for serine does not allow one of the interactions that rigidify Arg166. Thus this side chain is freer to move and expel the phosphate ion more expediently. Also, we learned that Asp153 participates in holding the Mg^{++} bound to the enzyme; replacement of Asp153 by the amino acid glycine (Asp153→Gly), which cannot hold the ion results is an enzyme that looses Mg^{++} readily.[3-4] The crying crystals of Chapter 5 were obtained, using this mutant. Stretching the metaphor, the crystals cry because the proteins used to grow them do cry (**Plate 21.1**).

The potential of genetic engineering combined with protein crystallography is enormous, and "protein design" is a thriving field of investigation. From a manufacturing standpoint, the attraction is that the bacteria produce the product. Genetic engineers are not synthetic chemists. They use the versatility and ingenuity of the twenty amino acids normally found in the cell to design and produce novel macromolecular entities. These molecules may have applications in medicine, diagnostics, ecologically friendly manufacturing protocols, processes of bio-remediation of contaminated environments, and other still unforeseen usages. The field has reached a point where attempts are being made to design proteins "*de novo*," with no pre-existing scaffold in nature.[5-9] The future looks very promising indeed.

Recent advances in the field of molecular genetics now permit not only the generation of recombinant bacteria, but also the possibility of altering the genetic make-up of organisms. These are referred to as transgenic animals or plants, and exhibit properties not present in their natural populations. For instance, transgenic mice are generated by injection of a recombinant DNA construct molecule directly into the male pro-nucleus of a fertilized egg. The injected eggs are transferred to pseudo-pregnant females and the resulting pups are tested for the presence of the foreign DNA in their genetic material. This technology can be used to among other things to: i) analyze the effect of single gene deletions or disruptions (i.e., knock-out experiments) in biological processes;[10] ii) produce mouse models of certain pathological conditions in humans (for instance, arthritis and muscular dystrophy); iii) introduce into plants or animals the genes for desired proteins; and iv) use the transgenic organisms for protein production. Transgenic mice can now be ordered on the Internet from vendors or academic institutions using available transgenic mouse core resources or laboratories.

In agriculture, similar technology is being used to generate, among other things, pest and herbicide resistant crops, and plants containing genes that would make them less sensitive to salt water.[11] The number of possible applications of genetic engineering is unimaginable. Every nook and cranny of the biosphere could be affected by the promise of bigger, better, more efficient or convenient processes for humankind.

The three-dimensional structure of DNA proposed by Watson and Crick unveiled to the scientific community and the world at large a beautiful, harmonic, impeccable, and innocent structure. Further scientific and technological developments have allowed us to manipulate at will the content of the atomic libraries within those molecules, in a large number of organisms on Earth. We are now able to repair, rearrange, and insert into these unique molecules our wishes, our dreams, and even our follies.

Although apparently soft in manner and appearance, genetic engineers can modify our natural environment in subtle, but possibly more profound ways than the hard engineers ever could. They can produce new varieties of bacteria, plants, and organisms with beneficial traits for the human species: oil-digesting bacteria, pest-resistant plants, salt-liking plants, and animals that can mimic pathogenic conditions in humans. This represents an extraordinary power; a power unprecedented in the thousands of years that *Homo sapiens* has walked upon the Earth. It is tantamount to being able change our physical and biological environment not from the outside where it can always be erased or obliterated, but from the inside where it has the potential of being retained. By virtue of the technology they have created, genetic engineers are the direct heirs of the Promethean fire that made the human species able to control the energy of the atom. Their flame has now reached the domain of the biological sciences. In doing so, biology as a field of scientific research has lost the innocence of a younger age. Early in the development of the field, scientists involved in this research evaluated the implications of this endeavor and a temporary moratorium on these experiments was agreed on;[12] nowadays, the consensus is a cautious go-ahead. In view of this unsurpassed power somebody may ask:[13]

> [...] Where are the taloned birds
> endlessly tearing the entrails
> of those who stole sparks
> from atoms
> to make
> bombs?
>
> Where are the guardians
> to set the world right
> when mortals trespass
> in the realm of the gods? [...]

Those guardian angels are nowhere to be found. We have to look into ourselves. We have to look into our innermost ethical and moral selves, and decide as individuals, as societies, and in the end as the sisterhood and brotherhood of the human species what to do and what to refrain from doing. Protein crystals will continue to help us to design more efficient catalysts, better materials, more potent vaccines and drugs, but—in spite of their magical reputation—they will not shine any light on distinguishing between morally good and ethically dubious genetic engineering experiments.

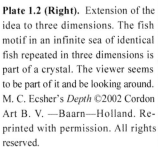

(110)
(130)
(120)
(100)
(010)

Plate 1.1 (Top). A two-dimensional crystal. The repetition of the same stamp along the horizontal (*x*) and vertical (*y*) directions is a simple example of how a two-dimensional crystal is created. Stamp design ©1984 United States Postal Service. Reproduced with permission. All rights reserved. Written authorization from USPS is required to use, reproduce, republish, upload, post, transmit, distribute or publicly display this image.

Plate 1.2 (Right). Extension of the idea to three dimensions. The fish motif in an infinite sea of identical fish repeated in three dimensions is part of a crystal. The viewer seems to be part of it and be looking around. M. C. Ecsher's *Depth* ©2002 Cordon Art B. V. —Baarn—Holland. Reprinted with permission. All rights reserved.

Plate 3.1. The "real" and "reciprocal" space construction using the two-dimensional crystal presented on Plate 1.1. On the original diagram, the distances of the short (horizontal) and long (vertical) spacings of the stamp were measured in centimeters (2.3 and 3.5 cm respectively; 1 inch = 2.54 cm). These values were inverted to give approximately 0.435 and 0.286 cm^{-1}. An arbitrary scale factor of 10 was applied to the reciprocal dimensions to bring the reciprocal lattice points (dark spots) to approximately the same surface. Notice, for instance, that the short spacing (1.9 cm) between the (110) planes results in the long distance from the origin to the 110 point of the reciprocal lattice of approximately 5.3 cm^{-1}. The short edge of the stamp in real space has been transformed into a long distance in the corresponding reciprocal lattice dimension (100) and the long edge into the short reciprocal lattice unit (010).

Plate 5.1. Collage of different protein crystals viewed and mounted in various ways. The two top views (**a** and **b**) depict the ways protein crystals are mounted to be exposed to X-rays. (**a**) Classical way inside a quartz capillary (notice the tiny speck in the middle between the "mother liquor" and the dark clay holding the capillary to the crystal holder. Right above the column of the mother liquor the capillary is sealed with wax to prevent drying. (**b**) Crystal mounted within a liquid cryo-protectant mixture supported by a rayon loop. Reprinted with permission from *Journal of Applied Crystallography*.[6] Courtesy of the International Union of Crystallography. The two bottom views (**c** and **d**) illustrate how the same two fresh crystals (**c**) of a mutant of the phosphate-cleaving enzyme alkaline phosphatase, develop cracks and fissures (**d**) upon the release of Mg^{++} ions contained inside the original crystal. Unless they contain chemical groups absorbing visible light, protein crystals are typically colorless. The color of some of the crystals in these images is due to the use of polarized light to make a more dramatic effect. The approximate dimensions of these crystals are 0.3–0.5 mm, or about one to two hundredths of an inch (1 inch = 25.4 mm).

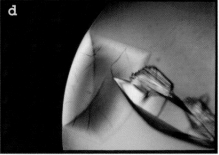

Plate 6.1. Symmetry in Space and Time. Superposition of a sample of the Moorish symmetrical patterns in the palace of the Alhambra, Granada, Spain, with the first page of the musical score of *Recuerdos de la Alhambra* by Francisco Tárrega Eixea. The music copyright ©1926 by Union Musical Española Editores, Madrid, Spain. International copyright secured. All rights reserved. Reprinted by permission of Associated Music Publishers, Inc. (BMI). C.A-Z appreciates the assistance of Jim Sukowski (Abbott Laboratories' Creative Network) in the composition of this image.

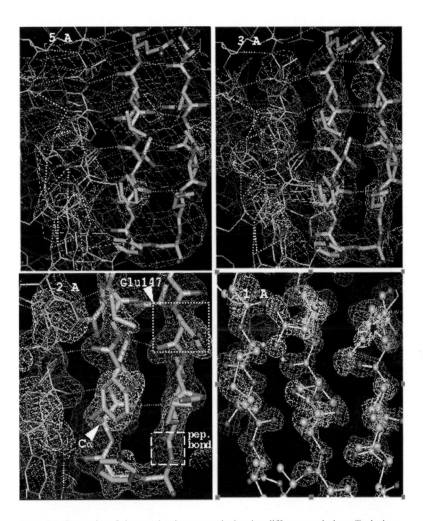

Plate 7.1. Examples of electron density maps calculated at different resolutions. Typical sections of electron density maps calculated at four different resolutions: 5 Å (2,160 reflections used); 3 Å (9,660); 2 Å (28,000); and 1 Å (105,000). The data for the 1 Å map calculation courtesy of Dr. V. Nienaber (previously of Abbott Laboratories). The final crystallographic structure has been added in all cases to show the "chemical interpretation" of the electron density. One amino acid unit (glutamic 147), the central carbon (C_α) and the peptide bond for different amino acids have also been outlined to aid in the interpretation.

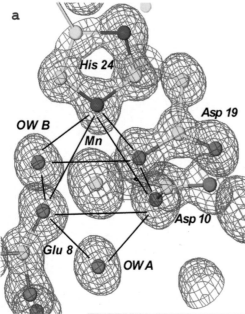

Plate 7.2. Two examples of electron density maps at very high resolution displaying individual atoms. Using the X-rays produced by synchrotron sources, crystallographic data of superb quality and high resolution can be routinely obtained for a larger number of proteins. The vast amounts of data permit the reconstruction of proteins with amazing detail (very high resolution), where the individual atoms making up these macromolecules can be resolved as individual spheres (a) Manganese ion (Mn^{++}) coordination sphere in the plant protein Concanavalin A (0.9 Å; courtesy of John Helliwell and co-workers.) The six atoms in approximately octahedral geometry around the Mn^{++} have been highlighted. (b) A detailed view of the intricate labyrinth of interactions within the atomic structure of human aldose reductase at 0.66 Å resolution (Courtesy of A. Podjarny, A. Joachimiak, and co-workers). A region containing amino acids tyrosine 48 (Tyr48), histidine 110 (His110), tryptophan 111 (Trp111) and a chemical entity named IDD594 is shown. Notice how the size of the spheres matches the size (number of electrons) of the different atoms: C, white, 6; N, blue, 7; O, dark purple, 8. At the top of the image W1 represents a water molecule interacting with a nitrogen atom in His110 via a hydrogen atom (i.e., forming a hydrogen bond). Amazingly enough there is a blob of electron density for this atom that contains only one electron within its electronic cloud. Numbers correspond to the distances in Ångstroms between the atoms.

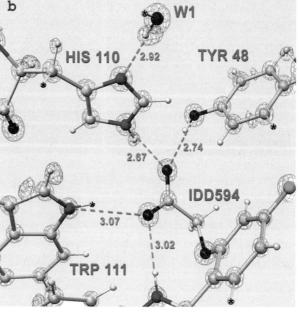

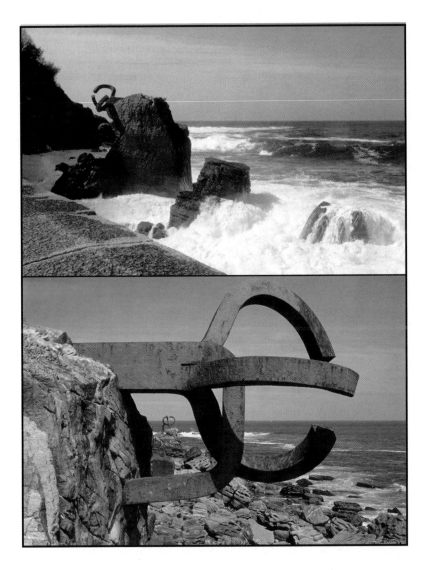

Plate 9.1. Collage of two different views of "The Combs of the Wind" with the Cantabrian sea in the background. I must thank Quintas Fotógrafos (Vitoria, Spain)—top, and my friend Ignacio Hernando—bottom— for the original images. The assistance of Jim Sukowski and Jeff Frye of Abbott Laboratories' Creative Network is appreciated.

Plate 10.1. The assembly of friends and colleagues of Prof. M. G. Rossmann on the occasion of the symposium entitled *New Directions in Protein Structure-Function Relationships* organized to commemorate his 65th birthday in front of the Lilly Hall of Life Sciences at Purdue University. West Lafayette, IN, USA, October 21, 1995. Many of the attendees to this celebration are prominent macromolecular crystallographers in their own right. A key and name list to the identity of the participants is provided on **Plate 10.1b** and **Figure 10.2** on page 79 respectively.

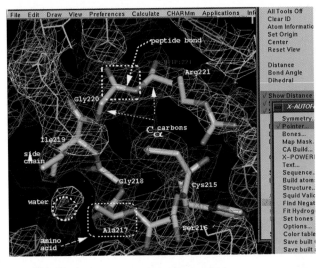

Plate 11.1. A multicolored screen from a modern-day crystallographic modeling program like QUANTA™ (Accelrys, Inc., San Diego, California). Diagram prepared with QUANTA™, captured with Snapshot™ and annotated with Showcase™. See explanations in the text, on pages 86–87.

Plate 12.1. The dramatic growth in number and complexity of the PDB holdings. Beginning with the 1970s, the first structures available to the scientific community included proteins such as myoglobin (**a**), hemoglobin (**b**), lysozyme (**c**), and the first transfer RNA (**d**). In the 1980s, advances in experimental and computational techniques permitted the solution of larger structures like antibodies (**e**) and the first icosahedral viruses (**f**). As of 2001, the combination of the widespread use of DNA recombinant techniques for protein production, and

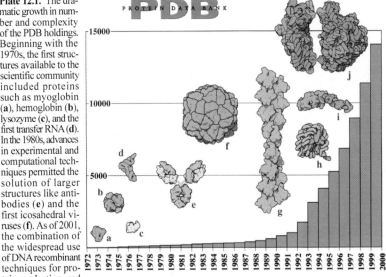

routine access to synchrotron radiation sources, has resulted in the elucidation of macromolecular structures of astonishing complexity. Illustrated are: actin (**g**) and myosin (**i**), muscle proteins; the nucleosome (**h**); and finally the ribosomal subunits (**j**). Structures depicted were taken from PDB entries: 1mbn, 2dbh, 2lyz, 4tna + 6tna, 1fcl + 1mcp, 2stv, 1atn, 1aoi, 1dfk, and 1ffk+1fka+1fjf. The image was created by Dr. David Goodsell of the Scripps Research Institute and is reproduced here with permission (©David S. Goodsell, 2001; Courtesy of David S. Goodsell.)[13]

Plate 12.2. Our modern, electronic, Plato's Caves. A human (yours truly) is looking at a molecular structure on the wall in a dark room, contemplating its beauty and wondering whether what he sees is real or just an abstraction or "ideal" in the Platonic sense.

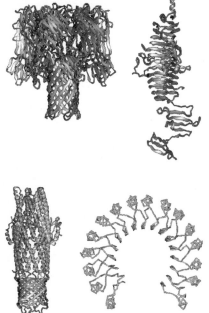

Plate 12.3. The artistry of protein structure. A collage of four beautiful structures from the PDB. **Top left:** A mushroom-like protein; alpha-hemolysin or alphatoxin, a pore-forming, cytolytic protein (PDB accession—not "ascension"—code: 7ahl). **Top right:** A protein suggesting the form of a hippocampus (sea horse); this protein is a part of the tail-spike protein from the bacteria-eating virus phage 22 (P22, code 1qrb). **Bottom left:** TolC, a tube-forming protein involved in the expulsion of foreign substances in bacteria (1ek9). **Bottom right:** Horseshoe-shaped protein; this protein inhibits an enzyme called "ribonuclease" (because it cuts ribonucleic acid (RNA)) and contains a functional unit called "leucine-rich repeats" (2bnh), repeated sixteen-times in the structure. The crystallographers involved in theses structures and the corresponding references can be found in the indicated files in the PDB. The different colors on the two left images have been drawn to indicate that these two structures consist of several identical subunits organized together. The basket-like portion at the bottom inserts into the cell membrane. Images prepared with QUANTA™, using the indicated records from the PDB.

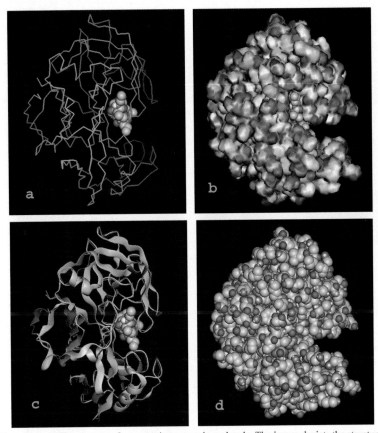

Plate 13.1. Different ways of representing a protein molecule. The image depicts the structure of a protein with a small molecule (an inhibitor) bound in the middle of the active site of the protein, blocking it. The **upper left** (**a**) shows what is referred to as a C_α representation, where a line connects the central carbon atom (C_α atom) of each amino acid. In the middle of the protein structure a cluster of solid spheres of different colors (white: carbon; blue: nitrogen; red: oxygen) is a representation of the small organic molecule that blocks the enzyme. The **lower left** (**c**) shows another schematic representation whereby the linear tracing of the **top left** (**a**) is represented by a ribbon. Regular parts of the structure known as β-sheets appear as a wavy ribbon and α-helices (as seen at the very bottom of the structure) as helical sections much like an old-fashioned telephone cord. Clearly this protein consists almost exclusively of β-sheets. The representations on the right emphasize the solid and atomic nature of protein structures. The **upper right** (**b**) illustrates "surface charge" with color-coding as in the solid spheres of the **lower right** (**d**). This representation gives an idea of the complex surface texture at the atomic scale of these macromolecules. The **lower right** (**d**) representation is the "real" atomic structure where every atom is replaced by a sphere of appropriate radius and color (charge; red: negative, oxygens; blue: positive, nitrogens). Notice the similarities between (**a**) and (**c**) for the linear representation and (**b**) and (**d**) for the solid one. The orientation is the same in all four images. The structure is from a protein-cutting enzyme secreted by the pathogenic fungus *Candida albicans* inhibited with a chemical compound designed and synthesized at Abbott Laboratories.[5]

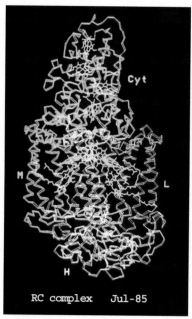

RC complex   Jul-85

Plate 14.1. The original image of the structure of the photosynthetic center of *Rhodopseudomonas viridis*. H, M, L, and Cyt label the four different components of the complex: Heavy (purple), Medium (blue), Light (orange) subunits, and Cytochrome (light blue). The edges of M and L present a vertical wall of helices, which forms the attachment surface for the cell membrane. Enmeshed in the scaffold of the polypeptide chain one can recognize the small-molecule components (yellow) that are critical for the light-trapping function of this molecule. Reprinted from *Nature*.[13] Courtesy of Dr. Hans Deisenhofer and Macmillan Publishers Ltd.

Plate 14.2. Rhodopsin from bovine retinal rod cells as the prototype of a G- Protein Coupled Receptor. Three different views are shown of the three-dimensional structure of this protein. **Top** view (**a**) is a special image known in the field as a stereo image or stereo pair. The two images represent two views of the molecule rotated about the vertical axis by a few degrees. Looking at each individual rendering separately with each eye results (after a few seconds) in a three-dimensional image of the object. Roman numerals label the seven trans-membrane helices characteristic of this class of membrane proteins. Reprinted with permission from *Science*.[14] ©2000 American Association for the Advancement of Science. Courtesy of R. Stenkamp, his co-workers, and AAAS.

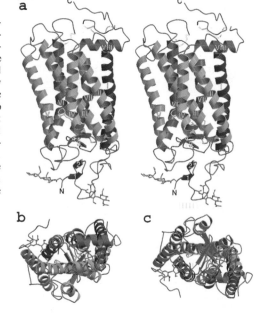

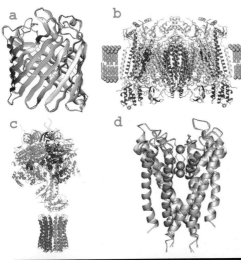

Plate 14.3. A collage of four membrane-bound proteins whose three-dimensional structures have been determined recently by protein crystallographers. (a) Porin.[15] Image created using PDB coordinate set 1opf; (b) Cytochrome *c* oxidase, image courtesy of T. Tsukihara and co-workers;[7] (c) F1c10 complex, the largest portion of the ATP-synthase known to near atomic resolution to date. Compound image courtesy of D. A. Stock, A. G. W. Leslie, and J. E. Walker.[8,9] The structure of the c10 fragment was obtained by Nuclear Magnetic Resonance methods by Mark Girvin and co-workers.[12] (d) Structure of the Potassium channel.[16] Image created with PDB set 1bl8. Images (a) and (d) created on a Silicon Graphics Octane™ with QUANTA™ (Accelrys, Inc., San Diego, California.)

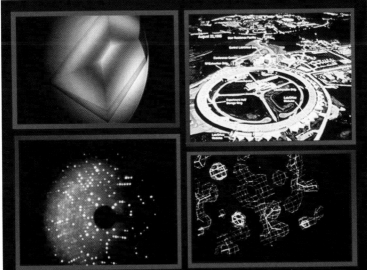

Plate 15.1. Structural analysis of macromolecules using synchrotron radiation. Proteins or macromolecules of biomedical interest are cloned, expressed and produced in milligram quantities using DNA recombinant technology. Samples are crystallized (**top left**), exposed to X-rays and complete diffraction data collected (**bottom left**, courtesy of Ed Westbrook). The availability of tunable X-ray sources of high brilliance such as the APS (**top right**) accelerates dramatically the process of structure determination of biological macromolecules of various sizes (**bottom right**, map at 1.8 Å resolution). The knowledge of the three-dimensional structures of protein targets is extremely valuable to design more efficacious drugs and to engineer proteins with specific and novel catalytic properties. Photo credits: Abbott Laboratories group.[19]

APS Collaborative Access Teams by Sector & Discipline

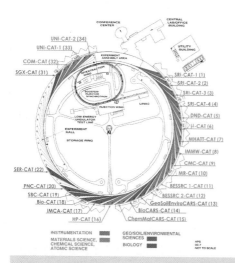

Plate 15.2. Schematic of the APS with the location of the different Collaborative Access Teams (CATs). The lines tangential to the storage ring represent the direction of the X-rays emitted by the cycling electrons. The approximate location of the different CATs around the ring is shown, clustered by sector and discipline. Courtesy of R. Fenner (Advanced Photon Source Public Affairs office). The wide range of disciplines making use of synchrotron radiation nowadays should be apparent.

Plate 16.1. Three views of the City of Grenoble, France. The upper left image depicts the very same spot where the original draft of this essay was prepared in the *Place de St. André* with a view of the *Palais de Justice*. The right image is a view of the banks of the river *L'Isère* with the *Vercors* mountain in the background. The rivers *L'Isère* and *Le Drac* can be seen in Plate 18.1 where an aerial view of the **ESRF** is shown in the upper right panel. One of the city gardens is shown in the lower left image. Photos courtesy of Ms. Chantal Argoud and Dominique Cornuéjols of the ESRF Public Relations office.

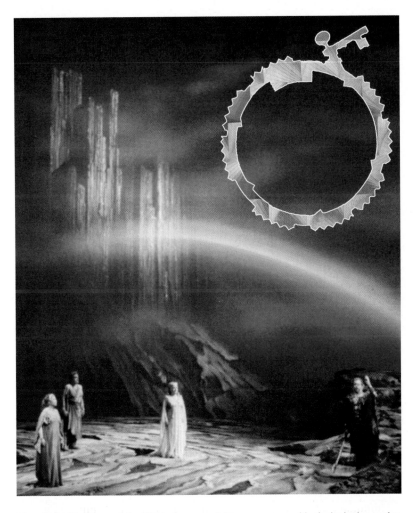

Plate 17.1. The image of the APS in the upper right corner appeared in the invitation card to the dedication of the Advanced Photon Source on May 1, 1996 (Courtesy of Argonne National Laboratory, Susan Barr Strasser). This ring-like representation inspired this chapter. The background illustration is reproduced with permission from a photograph entitled *Das Rheingold* Ms. Beatriz Schiller (Copyright ©2001 Beatriz Schiller) of the 1990 production of *The Ring* by the Metropolitan Opera of New York, James Levine conducting. The image captures Wotan holding the gold ring.

Plate18. 1. Collage of Cathedrals and Synchrotrons. (**a**) The current status of the Temple of *La Sagrada Familia* in Barcelona, Spain, and the "vision" for the reconstruction of the *Frauenkirche* in Dresden, Germany (**d**). On opposite sides, aerial views of the European Synchrotron Radiation Facility (ESRF) in Grenoble, France (**b**), and of the Advanced Photon Source (APS) in Chicago, IL, USA (**c**).
(**a**) Aerial photograph by J. Todó ©TAVISA, reprinted with permision. (**b**) Courtesy of the ESRF public relations office (**c**) courtesy of the APS users office and public relations. (**d**) ©1999 Blackwell Wissenchafts-Verlag GmbH, Berlin. Reprinted with permission. This image appeared on the Cover of *Mycoses* 1999, Vol.42, no.3. (Special Issue of 5th Congress of European Confederation of Medical Mycology) is entitled *"Wiederaufbau der Frauenkirche in Dresden: Fotomontage des Kuntslers und Fotographen Jorg Schoner."*[10]

Plate 19.1. The image presents an overall tracing (like in Plate 13.1c) of two subunits (dimer: blue, and tracings: red) of the Human Immunodeficiency Virus (HIV or AIDS protease. The axis of duplication (i.e., symmetry axis) runs vertical along the middle of the image in between the individual subunits. For emphasis, all the atoms and bonds of the therapeutic compound (i.e., drug) are depicted; green: carbon; yellow: sulfur; red: oxygen, and blue: nitrogen. The duplicated aspartate residues (yellow) in the two identical chains are also highlighted extending from symmetrical hairpin loops within the protein active site. The drug occupies the position of the molecule upon which the HIV protease acts on. The presence of the therapeutic entity blocks the activity of the enzyme and thereby prevents the HIV virus from growing and developing.

Plate 20.1. Molecular graphics rendition of an α-helix. Using the standard color-coding scheme (oxygen: red; nitrogen: blue; carbon: white; sulfur: yellow,) **(a)** shows the atomic structure of the α-helix. A different representation is used in **(b)** that emphasizes the bonds between the atoms. The dashed lines indicate the hydrogen-bond interactions (between C=O and H–N) that give stability to this structure almost perfectly aligned with the helical axis (vertical). **(c)** The outline of a helical envelope has been drawn using computer graphics. The side chains of different amino acids (for example, the nitrogen-containing arginine at the very bottom) are shown extending out from the helical surface. Of note is the long, sulfur-containing, amino acid methionine (top left) that can be replaced by the selenium-containing analog to facilitate the structure analysis of proteins using the **MAD** method.

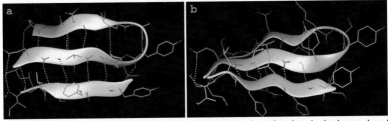

Plate 20.2. Two views of a β-sheet structure. **(a)** A top view showing the hydrogen bond (dashed lines) approximately in the plane of the paper and the side chains of the amino acids protruding out from the wave surface that characterizes the β-sheet. **(b)** The same structure, looking from the side. Notice the side chains of the amino acids alternating "up" and "down" from the plane of the wavy polypeptide chain that forms the β-sheet. Diagrams prepared with QUANTA™.

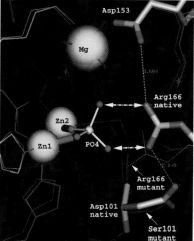

Plate 21.1. Close-up of the active site of the native and engineered *E. coli* alkaline phosphatases showing the critical residues for catalysis. The dashed lines between aspartic101 (Asp101, bottom), aspartic153 (Asp153, top), and arginine166 (Arg166) indicate the interactions that held the arginine166 side chain in a more rigid position (the distances of approximately 3.6 and 2.8 are in Ångstroms). Replacing Asp101 by Ser gives more flexibility to Arg166 and changes the catalytic properties of the mutant enzyme (see text, p. 165). Zn1, Zn2, and Mg mark the position of the three metal ions (two Zn^{++}, and Mg^{++}) that are part of the active site of alkaline phosphatase. Coordinates for the refined native enzyme from the PDB entry1alk (in thick bonds) deposited by E. Kim and H. W. Wyckoff.[14] The coordinates for the mutant Asp101Ser mutant are in thin lines.[3] Image prepared with QUANTA™ (Accelrys, Inc., San Diego, California).

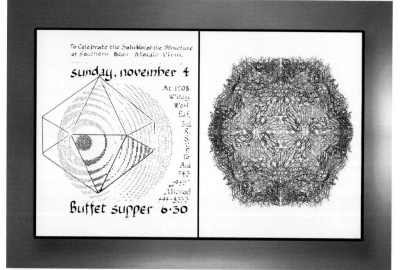

Plate 22.1. Left. Collage of the flier that Audrey Rossmann prepared in calligraphy as an invitation to the party in celebration of the structure of SBMV. The model made of small plastic triangles (in **Figure 22.1, right**) represents the arrangement of subunits in a virus like SBMV (T = 3). This image can be related directly to the symmetry in a soccer ball (**Figure 22.1**). **Right.** In the same orientation, the atomic model of the virus (Courtesy of my friend from our Purdue days, Keichi Fukuyama). The two virus particles are presented in the same orientation to draw attention to the main symmetry elements of the virus capsid illustrated in **Figure 22.1**: 2(green:gray)-, 3(blue:black)-, 5(red:white)-fold axes of symmetry. The five-fold clusters correspond to the black pentagons on the soccer ball. Notice that at the three-fold clusters six, almost identical, triangles meet: three green and three blue; quasi-hexagonal symmetry.

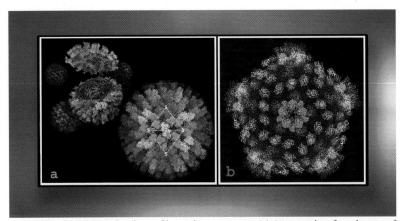

Plate 22.2. Two spectacular views of large virus structures. (**a**) A composite of two images of Blue-Tongue Virus. Reprinted from *Nature*.[6] Courtesy of David Stuart and co-workers (Oxford University, UK) and Macmillan Publishers Ltd. The symmetry elements of the icosahedral symmetry of these viral particles (5-folds, 3-folds, and 2-folds) have been labeled outlining the boundaries of the icosahedral asymmetric unit discussed before. The gray helical structures are intended to depict the nucleic acid. P, Q, R, S, and T label the different protein chains.(**b**) A view of the reovirus core. Reprinted from *Nature*.[7] Courtesy Steve Harrison and co-workers (Harvard University) and Macmillan Publishers Ltd.

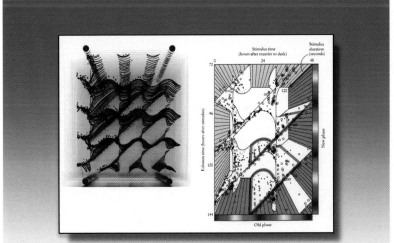

Plate 24.1. Left. Photograph of the original Perspex model of the time crystal. Old phase increases from left to right, stimulus duration from the back to the front and the new phase increases upward. **Right.** A more colorful representation of the same time crystal. Horizontal and vertical axes run as in the left but the stimulus duration increases from the foreground to the background as indicated by the red arrowhead in the top right hand side of the diagram. The helical boundaries of a pair of unit cells of the time crystal have been outlined in color. Reprinted with permission from reference,[2] p. 90. Courtesy of Arthur T. Winfree.

Plate 24.2. A larger, more artistic, rendition of the time crystal illustrating the two-fold screw symmetry (2_1) (rotation by 180° and translation upwards) of the surface making the unit cell of the time crystal. The axis of rotation is the singularity point. Reprinted with permission from reference,[2] p. 94. Courtesy of Arthur T. Winfree.

Plate 25.1. An illustration of the dynamic patterns of the BZ reaction. Note the symmetrical spiral patterns. Reprinted with permission from *The Timing of Biological Clocks* (1987).[14] Courtesy of A.T. Winfree and Scientific American Library. There are many recipes to prepare this famous chemical reaction. Several are given at the end of this reference book and in the publications of A.T. Winfree; more are given in reference.[6]

22

The Ballad of the 2.8 Å Structure of SBMV: Virus Structure

Most people would associate the term "ballad" with past achievements that could go back as far as the origins of story telling. It is certainly unusual to read or even hear this word associated with contemporary scientific events; even more so for the solution of the three-dimensional structure of a macromolecule. New structures are so commonplace nowadays. The major scientific journals of practically all biochemical fields are inundated with beautiful color pictures advertising on their cover the solution of yet another protein structure.

However, in the early 1970s a modern research project in structural biology reminded me of the mighty feats of medieval heroes. For nine long years, several generations of valiant postdoctoral investigators, led by Prof. M.G. Replacement, struggled to adapt the methodology of protein crystallography to the solution of the atomic structure of the first spherical virus particle. Viruses (from Latin *virus*, poison) are cellular parasites unable to reproduce by themselves. The class of spherical (or isometric) viruses was established to differentiate them from other plant viruses such as tobacco mosaic virus (**TMV**) that were more elongated, or fiber-like and generally helical.

There were two other groups working on a similar endeavor. One at Harvard led by Prof. Stephen Harrison, focused on Tomato Bushy Stunt Virus (**TBSV**), at the time the best structurally characterized spherical virus. A second group in Uppsala, Sweden, was trying to solve the structure of Satellite Tobacco Necrosis Virus (**STNV**), a very small satellite virus, under the *aegis* of Prof. Bror Strandberg.

The achievement established new methodology that is now routinely used, and the results obtained opened intriguing lines of research on the structure, function, and evolution of viruses. As I worked on the project, the stanzas of a ballad came to my mind as the most natural way to express my admiration for the feat of all the participants. It was a way of documenting the main milestones of a journey of scientific discovery of which I was a lucky participant. If you do not hear from the other groups, it is not because their achievement was less significant. Absolutely not; they simply could not find their *balladeer*.

The virus of our story is Southern Bean Mosaic Virus (SBMV), a humble RNA-containing plant virus that infects bean plants in the South of the United States. Neither SMBV nor its relative TBSV were ever as famous as the animal viruses that are fashionable today as human pathogens (for instance, AIDS virus, or rhinovirus, which causes the common cold). Small as viruses go (approx. 300 Å in diameter), non-enveloped, single-stranded, RNA plant viruses like them were easy to obtain in gram quantities from a few infected plants. In addition, they were easy to crystallize and consequently they were the objects of a concerted effort to obtain their atomic structure by X-ray diffraction methods with conventional in-house X-ray sources.

The icosahedral symmetry of small spherical viruses had been proposed by Watson and Crick in the early 1950s.[1] They hypothesized that the limited size of the coding material in a virus (either DNA or RNA), could only code for a few proteins and therefore the virus particle had to be assembled by the repetition of a single—or a few—protein chains. Therefore, they predicted that the virus envelopes would be highly symmetrical, and most likely icosahedral, containing at least twenty copies of the coat protein in the shell, or capsid. The detailed arrangement of the proteins in the capsid on the surface, and a preliminary classification of icosahedral viruses was presented by Caspar and Klug in their classic 1962 paper.[2]

It is instructive to look at the geometrical arrangement of protein chains on the surface of a virus particle from the viewpoint of more familiar objects. Think of a mosaic tile of hexagonal entities covering a planar surface. A moment's thought would make you realize that in order to curve a planar surface covered by hexagons you have to introduce a certain number of pentagonal links or tiles; otherwise the surface will always be flat (**Figure 22.1**, center inset). A close look at a soccer ball composed of pentagonal and almost perfectly hexagonal parts (quasi-hexagonal in the jargon of virus crystallography, Q6) sewn together illustrates the concept (**Figure 22.1**, left). Similar geometrical principles were used to build the celebrated geodesic domes of Richard Buckminster Fuller (1895–1983), and have now been seen again in a special form of carbon containing clusters of sixty atoms (referred to as fullerenes or buckyballs). In fact, it was recognized by Caspar and Klug that the inspiration for their insight into virus structure sprang from the arrangement of components in the geodesic domes. Based on this conceptual framework, the geometric lattices underlying the construction of certain spherical virus permits their classification as T = 1 (icosahedron: 60 protein subunits, 20 × 3), T = 3 (180, larger virus containing 180 subunits, 60 × 3), T = 4 (still more complex lattices containing 240 (60 × 4) protein subunits), and others (**Figure 22.1**, right; and **Plate 22.1**, right). The T (trian-gulation) number is a key parameter in understanding the three-dimensional structure and external appearance of icosahedral viruses.[2]

Nevertheless, there was yet no atomic model for an icosahedral virus particle well into the1970s. As initially proposed by M. G. R., the crucial factor in the determination of the structure was the presence of several identical copies of the polypeptide chain in the asymmetric unit: i.e., the presence of non-crystallographic symmetry. In those days, the major hurdle was to devise algorithms and programs, which would allow averaging of enormous electron density maps,

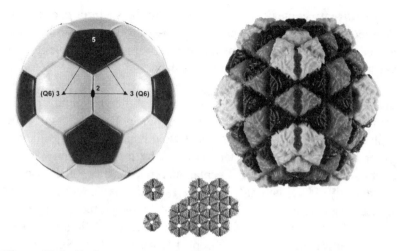

Figure 22.1. Surface structure of a T = 3 virus particle related to the external appearance of a soccer ball. On the left, the key symmetry elements of a virus are annotated on the surface of a soccer ball. 3 (Q6) identify the strict 3-fold axes of symmetry that are almost (quasi) hexagonal. The six edges of the hexagon are not quite identical; they only match in pairs. The center oval marks the exact 2-fold axis of symmetry and at the top the clear five on dark background identifies the conspicuous pentagonal shape on a soccer ball. These symmetry elements have their identical counterpart on the viral particle on the right. Note on the Q6 how two sets of three triangles (black and gray) meet. The triangle outlined on the left figure marks the boundaries of the unique fragment of an icosahedral figure. The remainder of the solid can be generated by applying the symmetry operations. Technically, that region is called the "icosahedral asymmetric unit." Compare with **Plate 22.1**.

containing many millions of grid points, over the redundant copies in the asymmetric unit. These methods were also employed to solve the structure of the common cold virus (rhinovirus) and are today standard in the structure determination of viruses and large protein complexes made up of several identical copies.

One of the most striking results of the structure determination of SBMV was the similarity of folds between the protein making the capsids of TBSV[3] and SBMV,[4] and later STNV[5] (**Figure 22.2**). This unifying principle had an enormous impact in understanding the structure, function, evolution, and diversity of a wide spectrum of viruses (**Figure 22.3**). Nowadays, the structure determination of a large virus is a dissertation project for a graduate student. A dramatic example of the spectacular results that virus crystallography has achieved are the structures of the core particles of two double-stranded RNA viruses. Blue Tongue Virus, solved by David Stuart and collaborators[6] and, most recently, the structure of the internal capsid particle of a different type of reovirus determined by K. Reinisch, Max Nibert, and Stephen Harrison from Harvard, form macromolecular assemblies of approximately 700 Å in diameter[7] (**Plate 22.2a,b**). Moreover, detailed structural analysis of virus structures is being used to understand how virus particles infect their host cells and to design drugs against the common rhinovirus and other viral pathogens.[8] The epic feat of the determination of the first atomic structures of viruses should be part of the folklore of macromolecular crystallography. Thus, it is appropriate that a full ballad was written to commemorate this colossal achievement using the technology of the time (X-ray film as detectors, rotating anode sources, and computers with limited memory) by three groups independently. Someday I may have the time and space to print the entire ballad, including the music. For this occasion, I would like to include only a few stanzas to give the reader a sense of the innuendoes of the text and of its rhythm and flow.

M. G. R. began to work on small plant viruses soon after the structure of lactate dehydrogenase (LDH) had been published. He went to Uppsala, Sweden, for a sabbatical leave in the laboratory of Bror Strandberg.

> LDH has now been solved
> I must find something to do
> Rossmann fold has been proposed
> I'll take a sabbatical leave *(repeat)*
> And I'll look at the STNV.

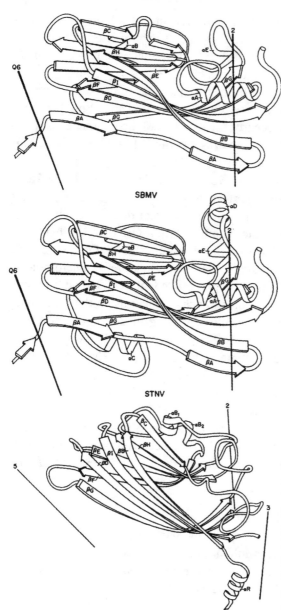

Figure 22.2.
Cartoon diagrams popularized by Jane Richardson of the polypeptide chains of the first three icosahedral virus structures obtained by X-ray diffraction on crystals.
Top: TBSV.
Middle: SBMV.
Bottom: smallest STNV.

The way the antiparallel β-strands wrap around the central core has been described as a "jelly-roll" or "β-barrel." The three structures have basically the same connections among the different strands labeled βA through βH. Lines mark the relative disposition of the symmetry elements of the icosahedron discussed in the text and in previous figures and color plates: 2-folds, 3-(Q6)-folds and 5-folds.[9] Reprinted with permission from Academic Press. ©Academic Press Ltd.

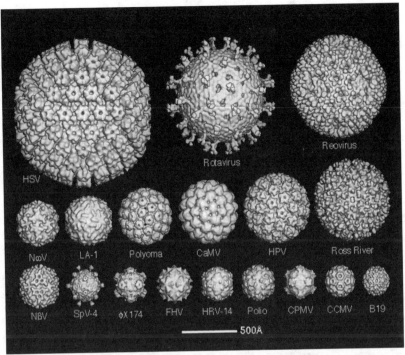

Labels in image: HSV, Rotavirus, Reovirus, NωV, LA-1, Polyoma, CaMV, HPV, Ross River, NBV, SpV-4, φX174, FHV, HRV-14, Polio, CPMV, CCMV, B19, 500Å

Figure 22.3. A collage of different virus particles. The image is a collage of the images of various viruses to give an idea of the wide range of sizes, shapes and textures present in these pathogens. HSV: herpes simplex virus; Rotavirus: agent of infantile gastroenteritis; Reovirus: any of a group of large widely distributed viruses, possibly tumorigenic (respiratory enteric orphan virus, 1959); NωV: *Nudaurelia capensis* omega virus; LA-1: *Saccharomyces cerevisiae* (yeast) L-A-1 virus; Polyoma: a virus found in rodents associated with various kinds of tumors; CaMV: cauliflower mosaic virus; HPV: human papilloma virus; Ross river virus; NβV: *Nudaurelia capensis* beta virus; SpV4: Spiroplasm phage 4; φX174: bacteriophage phi X174; FHV: Flockhouse virus; HRV-14: human rhinovirus 14, causative agent of the common cold; Polio: virus of the human poliomyelitis; CPMV: cowpea mosaic virus; CCMV: cowpea chlorotic mottle virus; B19: human parvovirus B19. The images are a representative gallery of reconstructions produced by cryo-electron microscopy (cryo-EM) in the group of Dr. Tim Baker (Purdue University). The atomic three-dimensional structure of many of these viruses has been determined by crystallographic methods. It would have been impossible to display such a variety in a single image using the atomic structures. The bar represents the distance of 500 Ångstroms.

Soon after the return from the sabbatical leave, he started to work on SMBV. Afternoon tea was a ritual at that time in the laboratory with a distinctly international flavor and where the work progress was discussed.

> Shall we start by growing some crystals?
> It's only a matter of weeks
> After that we can write some proposals
> For the future of SBMV *(repeat)*
> You should drink your afternoon tea.

Although M. G. R.'s dream was to solve viruses *ab initio*, he soon realized that heavy atom derivatives were a safer route at the time. Of course, he continued to sail in Lake Freeman, Indiana.

> Heavy-atoms must now be found
> Playing chemists is all we must do
> One alone will be safer ground
> For the structure of SBMV *(repeat)*
> I'm sailing the Indiana sea.

After eight years of work, the atomic model of SBMV slowly grew as a metallic sculpture made up of Kendrew parts in a forest of metal rods within a Richard's Box (**Figure 11.2**). M. G. R. kept bumping his head against the top of the box, so he purchased a hard hat, which he rigorously wore for his building sessions with me.

> Eight years have already passed
> Many people have done their best
> I won't say the struggle has finished
> For the structure of SBMV *(repeat)*
> I'll buy a helmet for me.

The fold of SBMV turned out to be a β-barrel (or jelly-roll) almost identical to the one already described by Steve Harrison and colleagues for the structure of TBSV (**Figure 22.2**).

After so many years of labor
All we have is a barrel of sheet
Old Steve did us a favor
With the structure of TBSV *(repeat)*
We can trace our SBMV.

The entire original text of the ballad was sung at a party at the Rossmann's residence on Sunday, November 4, 1979, to celebrate the solution of the three-dimensional structure of **SBMV** (**Plate 22.1**). The melody was adapted from a song by Pete Seeger (the grandfather of American folk music) that I heard on the radio one beautiful autumn morning on my way to the lab. The entire ballad was meant to be an homage to all participants in the project of the three-dimensional structure of SBMV. Many of them I have met through the years at meetings and conferences. I shared with others hours of effort, frustration, and excitement in the basement of the Lilly Hall of Life Sciences at Purdue University, West Lafayette, Indiana. The years spent in that laboratory still bring cherished memories. It is a unique laboratory whose day-to-day routine is still masterfully orchestrated by Sharon Wilder. To all of them (the unsung heroes of this ballad) and to many others who participated in less visible ways, I want to express my deep appreciation: Sherin Abdel-Meguid, Toshio Akimoto, J. E. (Jack) Johnson, Andrew G. W. Leslie, Ivan Rayment, Michael G. Rossmann, Ira Smiley, Dietrich Suck, Tomitake Tsukihara, and Mary Ann Wagner. I am just a modest minstrel, the troubadour of this epic feat.

VIII

future perspectives

23

The 1.8 Å Structure of Scientific Revolutions

As the reader might have already inferred, I have a philosophical slant in my interests and perceptions. Therefore, it will not come as a surprise that in my graduate student days at the University of Texas at Austin (under the mentorship of Profs. M. L. Hackert and J. L. Fox), I came across a copy of the nowadays classic book *The Structure of Scientific Revolutions* by T. S. Kuhn.[1] With my still crude English-reading skills, I read it from cover to cover and realized immediately that it was a milestone. Not only in understanding the way science was performed on a day-to-day basis, but also in putting into the right perspective the accumulated knowledge resulting from the labors of scientists and researchers in the different areas. I was a student of protein crystallography at the time and as a joke, I wrote on the blackboard the title of the book with "1.8 Å" inserted just before the word "Structure": The 1.8 Å Structure of Scientific Revolutions.

The first edition of Kuhn's classic was published in 1962 as a collection of brief essays analyzing the way science was made by its practitioners, and how new scientific theories emerged. In the same year, another work of related themes was published by M. Polanyi[2] where he uses an instructive metaphor to illustrate the inner works of the scientific activity. According to Polanyi, doing science is rather like putting together a gargantuan jigsaw puzzle. Bauer has explained this analogy in detail and elaborated further in a provocative book.[3]

I will sketch the basic ideas of the analogy to introduce some of the concepts in T. S. Kuhn's influential essay.

Several strategies are possible in building a jigsaw puzzle from scratch by several people. One can divide the number of pieces in equal numbers and distribute them among the participants. This approach will be very ineffective since few of the pieces allocated to each person would fit together. A slightly better method might be to complement each participant's lot with a copy of all the pieces and eventually combine all the results. Either way, the group as a whole would not perform much better than the individual partners working separately. The only way the group can effectively surpass significantly what any member can do alone is to let them work assembling the puzzle together, each colleague fully aware of what everybody else is doing. In this way, each participant will work on his (or her) own initiative and strengths, and will respond immediately to the latest accomplishments of the others. In a nutshell, independent initiatives and actions lead to a joint result, which is unpredicted by any of the individual participants.

This simple metaphor captures many of the characteristics of the scientific enterprise: It works best under a cooperative umbrella; it is a systematic effort, and it benefits from individual initiatives. It can also be used to illustrate three basic concepts introduced by Kuhn: "normal science," "paradigm," and "scientific revolutions."

Puzzle-builders involved in filling the gaps of a section of the puzzle that is clearly outlined will be doing normal science; no real surprises are expected and it is only a matter of time before that section is completed. A person or group of colleagues suggesting that an unknown area of the puzzle resembles a forest, where many previously abandoned pieces can now fit, would propose a paradigm. By considering not only the color but also the texture of the component pieces, another person or group notices that the forest consists of spruce trees and other unexplained features, giving rise to a scientific revolution. This crisis could end up by realizing that the forest was indeed part of a mountain landscape, resulting in a "paradigm shift". What were before a few odd and inconsistent "edge" pieces within the forest paradigm can now be fitted perfectly into the new and expanded framework. These concepts created quite a controversy in their time, which in certain circles still rages on—especially because they called into question the concept of scientific method and its role in the practice of science.[3]

I would like to extract a brief quote from Kuhn's analysis to point out how well his informal and completely unrelated definition of a "paradigm" fits the determination of the three-dimensional structure of the first proteins by Kendrew, Perutz, and their co-workers, a feat that created the discipline of protein crystallography and illuminated a new way of looking at the molecular components of life processes:[4]

> "[...] Their achievement was sufficiently unprecedented to attract an enduring group of adherents away from competing modes of scientific activity. Simultaneously, it was sufficiently open-ended to leave all sorts of problems for the redefined group of practitioners to resolve.
>
> Achievements that share these two characteristics I shall henceforth refer to as 'paradigms', a term that relates closely to 'normal science'. [...]"

Nonetheless, the content of Kuhn's book lay dormant for many years in some arcane crevices of my brain. Over the past few years, I have come to reflect again over its concepts in relation to the so-called revolution in the biological sciences originated by: i) the determination of the three-dimensional structures of DNA and proteins at atomic resolution; and ii) the discovery of the Genetic Code and the advances in understanding the machinery of the cell at the molecular level.[5] The common use of DNA recombinant technology and the availability of multitude of protein structures as determined by X-ray diffraction and NMR methods can be taken as tangible proof of such a revolution. Molecular and structural biology are today commonly accepted concepts as well as the subjacent notion that the atomic structure of biological molecules can explain their function.

As practitioners of the trade that is responsible for the majority of those atomic masterpieces, crystallographers have been very successful in producing structures of ever increasing complexity and size in amazingly short periods of time. There is an informal record circulating in the community of three months from an isolated target gene to a complete three-dimensional structure determination of the associated protein product, well-refined and with excellent structural parameters. The time of the record is probably shrinking down dramatically as you read these lines.

As discussed in previous essays, the future looks even brighter with the availability of third generation synchrotron sources (ESRF, APS, and SPring8), CCD detectors and almost unlimited computing power, combined with superb software tools that our community has developed over the years. Soon, we will be able to slice biochemical and enzymatic events at the microsecond or even nanosecond time scale.

Our results and methods are being used to expedite and rationalize the design and development of new drugs and diagnostic tests for the patient community at large, and enzymes with novel properties are being engineered by random and directed mutations of the three-dimensional structure of the wild-type enzymes.

And yet, as conscientious professionals we also realize that our tools and methods have limitations such as: i) many interesting target proteins do not seem to crystallize in our laboratories or may never be crystallized; ii) the quality and extent of the diffraction patterns from our crystals never seems to be adequate to get answers to many of the interesting biochemical or biological questions; iii) our electron density maps, and consequently our refined protein structures have disordered regions that cannot be determined unambiguously; and iv) our refined models have only limited information regarding the mobility and dynamics of the macromolecular structures we work on.

In the domains of structure, flexibility, and dynamics of macromolecules in solution, the NMR community has also made tremendous strides in adding, expanding, and complementing the knowledge derived from X-ray diffraction. They also face their intrinsic limitations and challenges and what their ingenuity and efforts will yield in the future can only be a matter of conjecture.

Through our education, we have all been exposed to the limitations imposed in our knowledge of the atomic world by the Heisenberg's uncertainty principle governing the simultaneous determination of the position and momentum of an electron in motion. Of course, you will argue, our results do not correspond to this domain of knowledge, but some of the work on the migration of carbon monoxide in myoglobin,[6] or the electron paths in electron transfer processes[7,8] may point in that direction. Could it be that the limitations listed above will impose an analogous uncertainty principle for our atomic description of the biochemical phenomena responsible for some of life processes? Prof. R. Huber mentioned this idea to me almost thirty years ago in an informal conversation in St. Paul, Minnesota. I do not mean to imply the existence of an external principle, an *"elan vital,"* but only that our methods and probes into the atomic structure of the biological matter have limitations. These limitations may come in two separate domains: the structural and the functional. First, crystallography may be unable to provide the necessary atomic (static) detail to describe a certain biochemical process. Second, biological processes are inherently dynamic phenomena, and crystallography has only a limited power to analyze the flows and fluxes, upon which life is based on.

The efforts of protein crystallographers, individually and as a community, have undoubtedly resulted in an on-going scientific revolution. The instruments of this revolution are the tools that the community has developed to examine the atomic structure of biological macromolecules aided by the dedicated efforts of molecular biologists and protein biochemists, supporting the work, and complemented by the tools that the NMR community has developed on its own.

The icons of this revolution are the myriad of well-refined three-dimensional structures of macromolecules of paramount biological importance displayed in various scientific journals, many of them well beyond the ones dedicated to crystallographic work. These icons are deposited in the Protein Data Bank and have already made their appearance in the current textbooks of Biology, Biochemistry, and Medicine: they are already part of what T.S. Kuhn would call "normal science", distant already from the "revolutionary science" that we pioneered only a few years ago. Thus we must admit that like any other scientific or political revolution, ours is also limited in extent and scope. Whether this limitation is inherent to nature or only restricted by the tools we currently use, it is still a matter for the future to decide.

24

A Crystal in Time:
Biological Clocks

xcept for the time that takes to solve our crystal structures, the variable "time" does not play an important role in the professional life of protein crystallographers. The variables, upon which we concentrate all of our efforts, are the spatial coordinates either within our electron density maps or as the triads (x,y,z), placing the atoms of our chemical models in space. As many other people have argued before, our science and our results are fundamentally static. Only the temperature factors (or B factors) associated with atoms or groups of atoms in the crystal give a glimpse of the incessant motion of our atomic universe.

During my post-doctoral years at Purdue University in West Lafayette, Indiana, I was fortunate to meet and become a friends of a very special person whose main interest was time and its relation to the study of biological clocks. At that time, Arthur T. Winfree was a well-known and respected figure in the field and had just published a major book entitled *The Geometry of Biological Time*.[1] The monograph was an amazing compendium of observations and mathematical models of what was known at the time about biological clocks. We used to have lunch together at some of the local eateries in West Lafayette, Indiana, and these social encounters were my first introduction to the fascinating world of the circadian regularities in living organisms. Unfortunately for me, he left Purdue University for warmer climates soon after my arrival, and as a good-bye present he gave a copy of his monograph with the following dedication:

"For CAZ, crystallographer of space
From ATW, crystallographer of time"

I was very intrigued by those few words and inspired by Arthur's personality and approach to science. Dr. Winfree went on to gain recognition for his iconoclastic and imaginative research in cardiac arrhythmias and was awarded a well-deserved J. D. and Catherine T. MacArthur Prize. During the ensuing postdoctoral years at Purdue and during my non-existent spare time, I read some sections and tried to grasp the fundamentals of the field. Naturally I failed, but a few years later I rediscovered the theme in a simpler, more descriptive, and beautifully illustrated, version of the original monograph entitled *The Timing of Biological Clocks*.[2] In its new incarnation, the universe of circadian rhythms swallowed me for about two to three months and even though I have certainly not mastered the field, during my readings I discovered a fundamental theme that percolates through the biological clocks of many living systems. This fundamental observation was baptized by Arthur T. Winfree as the "Time Crystal" and was first discovered in a species of the fruit fly *Drosophila pseudoobscura*. In addition to the word crystal, there is also an anecdotal and historical connection to protein crystallographers. The data showing the first *time crystal* was plotted on Perspex sheets in Cambridge, England, in the same workshop where the pioneers of protein crystallography built stacks of electron density maps to visualize the early protein structures (**Plate 24.1**, left).

I am by no means an expert in the field of biological clocks but I'll try to introduce the basic concept of the *time crystal* to the crystallographic community for two reasons. First, as a small homage to another outstanding scientist and friend of mine in a field different from crystallography. Second, as an inspiration to the new generations of protein crystallographers. Nowadays, our trade has become so streamlined, some of the old timers might even say almost effortless: New structures are solved and refined at an ever faster and alarming rate such that perhaps the new generations are wondering why they got into protein crystallography in the first place. Now, I would like to point out to them that they should look for inspiration in solving problems related to the interface between our static structures and the quintessential dynamic processes of life. How do biological clocks work at the molecular level? What is the structure of their essential molecular components? How are they orchestrated together in three dimensions? How do the physico-chemical properties of the microscopic cellular milieu produce this circadian dance in so many living systems, ranging from the rhythmical tidal glow of *Gonyaulax* cells, to the hatching of a population of eggs in *Drosophila*, and to the collective rhythm of the flowers of the Kalanchoë plant?

The existence of an internal clock in many different biological systems with an approximate period of 24 hours (circadian: *circa*— approximately, *dies*—day) has been well-established (see for instance the two books mentioned). Of interest for the ensuing discussion is the fact that within the *pupal case* of the fruit fly *Drosophila* (rice-like structures where larvae await their flight to adulthood), the brain of the larva keeps time and dictates the emergence hour of each particular individual. In this state, the motionless pupa is a self-contained system, which does not exchange any food or excreta with its surrounding environment. In nature and in a laboratory that is exposed to a 24-hour cycle of equal days and nights, the emergence of the individual or eclosion occurs in the first hours of daylight. Typically, the adult individuals emerge from their pupal cases in bunches or bursts, the timing of which is a reflection of the threshold of their internal clocks.

The pupal cases do not exchange solids or liquids with the surrounding environment; their only link to the external medium are two breathing tubes. Yet, they are sensitive to external stimuli, specially light. Rearing the larvae under constant light suppresses the ticking clocks and it turns out that these clocks are also blind to red or even yellow light; yet they are extremely sensitive to blue light. The eclosion of an entire population of pupae can be put in synchrony experimentally by collecting them in bright fluorescent lights and them putting them in a red or yellow environment. However, even a brief exposure to a perturbing light penetrating their eternal darkness offsets the timing of all the subsequent bursts, as though the incoming photons had reset the original phase of the internal clock from its old value to a new phase. This new phase depends also on the intensity of the perturbing light.

I apologize for the lengthy preparation but I could not explain the time crystal existing in biological clocks without introducing the identity and meaning of the three axes: x-axis (horizontal), old phase of the circadian clock (hours); y-axis (vertical), new phase (hours); and z-axis (into the page), stimulus duration in seconds from low to high. When Dr. Winfree plotted in perspex sheets the summary of several hundreds of experiments of perturbed eclosion events in *Drosophila* larvae, he found a repeated pattern that he labeled the time crystal (**Plate 24.1**). The three-dimensional plot showed how the new phase induced by the perturbation was dependent on the pre-existing—old—phase, and on the intensity (duration) of the external stimulus. The diagram displayed a 2_1 screw axis of symmetry (i.e., a 180° rotation, followed by a translation of half the full value) in the singularity point where the switch between odd and even resetting takes places (**Plate 24.2**). Time, space, and the limitations of my own knowledge prevent me from discussing the subtleties of this pattern that has been found in many other circadian clocks when the old phase is reset by different external stimuli to a new phase. I do encourage the reader to consult some of the details in the books that I have introduced.

Thus, crystalline symmetry is found not only in the geometrical patterns that we are so accustomed to in our everyday experience. It has also been unveiled in the internal machinery of dynamical processes that are essential to living systems. If I were to be a young macromolecular crystallographer again, it is within this domain that I would look for new scientific puzzles. It is the causal connection between our static structures and the rhythms of life that intrigue me.

Well, perhaps the new generations will move one step further in this direction, now that our friends, the molecular and cell biologists, have cut a trail towards some of the proteins responsible for these circadian clocks.[3] Moreover, the publication of the complete genome of *Drosophila melanogaster*[4] will undoubtedly provide new and intriguing clues into this fascinating problem. Fittingly, an expanded and updated second edition of the seminal book by Winfree just appeared with a list of approximately two hundred questions related to biological clocks that need to be answered.[5] The questions encompass cyclic systems from microorganisms to the "lunar message" of the female menstrual cycle. This field is indeed ripe for further exploration and elucidating the three-dimensional structure of the biomolecules involved in these processes will provide additional clues as to the understanding of their physico-chemical oscillations.

25

Conservative and Dissipative: Durable and Ephemeral Structures

t was a serene and beautiful evening in the fall of 1977. As I stepped out of the Patterson Laboratories building on the University of Texas at Austin campus, the bright orange lights coming from the windows of the tower in the Main Building surprised me. It was too early in the football season and having lived in the U.S. for a few years, I knew it could not be related to the results of the Southwestern Conference. So I asked. I was pleasantly surprised to hear that the reason was because Prof. Ilya Prigogine had been awarded the Nobel Prize for his work on non-equilibrium thermodynamics.

To me it was great news. Although not in my area of graduate studies, I had been attending Prof. Prigogine's informal seminars around the UT campus and even in my ignorance, I thought it was an interesting field of study to understand biological processes. Just months after my arrival in the U.S., I remember very vividly two informal seminars that he gave at lunchtime in our building (Patterson Laboratories) to summarize two articles that had appeared in *Physics Today* in 1972 on *The Thermodynamics of Life*.[1] During my years of graduate study on the UT campus, I saw Prof. Prigogine often at the local donut shops in the evenings and nights "burning" the midnight oil, like another graduate student. I still treasure my personal copies of several of his books with dedicatory autographs on them.

I think that my interest in this field of study stems from a lecture that our Professor of Biology at the University of Valladolid (Spain), Prof. José Planas Mestres, gave on the first day of class. He discussed how living systems violated the second law of thermodynamics because they grew and developed by decreasing entropy. Although an oversimplification and incorrect, this was a common notion at the time. It seemed like an introductory lecture and not so many students attended class that day. In the best tradition of the Spanish universities, this topic was the essay question of the term exam. Nevertheless, it was an intriguing question that has fascinated quite a few theoreticians, including Prof. Prigogine.

In spite of its name, the branch of physics that study thermodynamics is mainly concerned with a static system in full equilibrium with its surroundings. Processes and systems are always studied in their reversible limit and if time permits, one lecture is presented at the end of the year to discuss non-equilibrium systems. It was the extension of the concepts, methods, and theories of equilibrium thermodynamics to a more dynamic domain that has dominated Prof. Prigogine's lifetime scientific interest.

He was born in Russia and studied in Brussels where he became Professor (in 1951) and Director of the International Institutes of Chemistry and Physics. In the mid 1960s he taught at the Enrico Fermi Institute at the University of Chicago (1961–1966) and from 1967 had a double appointment between the Center for Statistical Mechanics at UT Austin and Brussels. In his early work on non-equilibrium thermodynamics, he was able to show that for systems that are near equilibrium and are able to exchange matter and energy with their surroundings (i.e., open systems), there was a "guiding principle" or direction towards minimum entropy production.[2] For example, this is the domain where the physico-chemical processes that allow the growth of crystals take place. However, a much more challenging and richer field of endeavor is the study of systems, which are maintained far from equilibrium such as living organisms, which are clearly able to organize themselves in spite of the universal tendency towards chaos and disorder. The term *"dissipative structures"* was first alluded to by R. Landauer in 1961[3,4] but I learned about it from the writings of Prof. Prigogine. In the 1960s he defined the term *"dissipative structures"* to refer to the internal organization generated in systems maintained far from the thermodynamic equilibrium by a constant flux of matter and energy from outside of the system. These structures can be "either temporal (chemical clocks and periodic oscillations) or spatial (Benard convection cells in fluid dynamics), or spatio-temporal (chemical waves)."[3] These ordered states are to be distinguished from our well-known crystals, which are equilibrium structures. Dissipative structures survive only due to the permanent constraints of the system with its surroundings, and the appropriate irreversible exchanges of matter and energy with the exterior of the system. In contrast, *"conservative structures"* are stable after their synthesis or generation and do not require any further energy input. I think that a few examples will clarify these two concepts.

Molecules such as proteins, DNA, and all biological biopolymers are conservative structures; once they are synthesized they do not require any more energy to retain its conformation. The best charac-terized example in the laboratory of dissipative structures are the spa-tial and temporal patterns produced by the Belousov-Zhabotinskii (BZ) reaction (**Plate 25.1**). This reaction was originally discovered by the Russian chemist B. P. Belousov while investigating the inter-mediates of a certain reaction involving the loss or gain of electrons (oxidation/reduction reaction). Specifically, he was investigating the oxidation of citric acid (the most abundant acid of lemon juice) by potassium bromate ($KBrO_3$), a potent oxidant. The reaction proceed-ed in acid media (with abundant H^+) in the presence of cerium irons (Ce^{3+}), which acted both as a catalyst and as a color indicator.

To his surprise and amazement, the initially homogeneous solution oscillated like a clockwork between a pale yellow and a colorless state.[5,6] A few years later, A. M. Zhabotinskii, a graduate student at

Moscow State University, reproduced the original results and was able to obtain similar oscillations replacing citric acid by malonic acid (another simple organic acid). News of the unusual behavior of this simple chemical reaction reached other laboratories in the Western world a few years later and the reaction was renamed Belousov-Zhabotinskii (or BZ) reaction. Since then, variations of the BZ reaction (i.e., replacing Ce^{3+} by iron and a dye makes the oscillation easier to visualize), are studied all over the world as a model system for oscillating reactions. In its simplest incarnation, the reaction is performed in a petri dish containing a shallow layer of solution; spinal patterns like the one presented in the figure appear gradually and soon fill the entire dish.

The BZ reaction and the unique spatial and temporal patterns that can be observed, produced, and sustained by its chemical ingredients have altered our view of chemistry. Agnes Babloyantz, one of the alumni of Prigogine's school of thought, has expressed it very lucidly:

"A chemical process is no longer only the mixing of some inert products to produce other 'lifeless' components. Under far from equilibrium conditions, a necessary condition for living species, a chemical reaction becomes 'alive.' It measures time, it propagates information, it discriminates between pasts and future, between left and right. It may also construct mosaic structures. If pushed too hard it starts to vacillate and behave erratically or 'irrationally'."[7]

If these simple chemicals, far from equilibrium, can present such an amazing oscillating behavior, what can we expect from the interactions among the richly textured structure of biological molecules? Isn't it conceivable that the molecular choreography that makes mitosis and meiosis possible is indeed an interconnected, dynamic, tapestry of related complex dissipative structures? A plethora of biological processes can be thought of as interconnected networks of these ephemeral events in the life cycle of bacteria and higher organisms alike.

Let us find some common day examples. Most of the things that we manufacture and handle in our daily lives are conservative structures: a book, a table, a car, etc.; although they eventually decay, once they are manufactured they do not require an influx of energy to retain their structure. However, the transient, almost ethereal things in life (i.e., flowers, a melody in an underground passage, a rainbow, a birdsong),[8] perhaps the things that we appreciate the most and indeed life itself are dissipative structures. For instance, a ballet and an opera performance are complex dissipative structures with the unique spatial and temporal structures created by the performers and the orchestra. Once their energy input ceases, so does the performance. Also, the momentous live performance of Beethoven's 9th *Choral Symphony* commemorating the fall of the Berlin Wall is a dissipative structure in time. Such a performance can be converted into a unique recording (a conservative structure on the grooves of the CD) by the magic of high-fidelity sound recording. Audiophiles know very well how ethereal those unique performances are, and are willing to comb exhaustively record stores and catalogs to find such a "collector's item."

For many years now, I have been fascinated by this dichotomy and especially by the interplay between the two concepts. It is clear that certain conservative structures allow us to create some dissipative structures in space and/or time and vice versa. The existence of a technology to manufacture a flute allows me to play a beautiful melody, given the right influx of wind and pressure on the keys. In turn, this melody may inspire us to write a poem or to create a sculpture, both durable structures. In the domain of sciences, I would venture that perhaps an "idea", or a flash of "inspiration", might very well be a transient spatio-temporal (dissipative) structure in our brain.[*] After applying the rules of logical knowledge and hypothesis testing, this idea might result in an invention, a theorem or a physical theory to explain some natural phenomenon. Undoubtedly, the evolution of human culture has been accelerated by our ability to convert or record transient effects (i.e., the spoken word, thoughts, an idea, or a melody) into

---

[*] I came to this conclusion on my own. The connection between thought processes and dissipative structures has been discussed also by Fritjof Capra in his 1996 book *The Web of Life: A New Scientific Understanding of Living Systems*.[9]

durable structures (written language, a technical sketch, or a musical score). Beginning with the initial abiotic synthesis of the first biopolymers (conservative structures), under the right environmental conditions a self-duplicating system developed (dissipative system). This, in turn, was able to synthesize and develop a more complex and efficient set of molecules (conservative structures), which gave rise to a larger variety of dissipative forms in space and time. This spiral of increasing complexity, alternating between the "conservative" and "dissipative" domains probably reflects the evolution of life in our planet (**Figure 25.1**), and has been the focus of detailed analysis at the molecular level, initially by M. Eigen (b. 1927).[10]

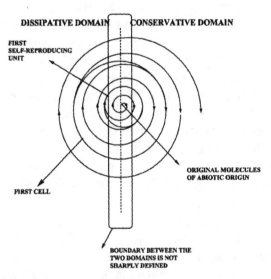

INTER-RELATION OF STRUCTURES

Figure 25.1. Inter-Relation between Conservative and Dissipative structures. An schematic representation of the interplay between conservative and dissipative structures in the evolution of complexity in the biosphere. The boundary between the two domains is probably not as clear cut as I suggested in the text and therefore the interface is not a line but an area or rectangular region. Circles in this spiral of ever expanding complexity could represent stable units such as the first self-replicating unit or the primordial cell. Diagram by the author.

In more recent years S. Kaufman has explored the implications of pre-existing complexity and order in biological evolution.[11] As I have discussed through the pages of these notes, trapping biological molecules into crystals is an excellent way to characterize them in three dimensions. Now, I wonder, will we ever be able to obtain glimpses of the internal order in space and time of dissipative structures? Perhaps, with our ever more intense, tunable, time-dependent synchrotron X-ray sources and ingenuous experimental conditions we might succeed. This will open another window of understanding into the inner works of living systems.

The ensemble of the myriad of molecules and macromolecules in an *E. coli* cell surrounded by the right milieu result in what we call "life". At the risk of being called a vitalist I would rather call it *"the living state,"* a state of matter different from the traditional solid, liquid, and gas. Not because I believe in the existence of an *"élan vital"*, but rather because it is the combination of the parts with the flows and the ebbs that these components make possible, that constitutes life. How all the parts interact to yield life is still the mystery but I am reasonably certain that the dynamics of dissipative structures and other non-linear phenomena are at the very core of it. The three-dimensional structures of all the macromolecules existing in a small bacterium, with their intricate complexity, will keep crystallographers and their latest technologies busy for a few years still. However, albeit important, they are only one part of the puzzle. The same thoughts have been expressed by R. Lewontin in a recent book (italics are my emphasis):[12]

> "Rather than searching for radically different ways of studying organisms or for new laws of nature that will be manifest in living beings, what biology needs to do to fulfill its program of understanding and manipulation is to take seriously what we already know to be true. It is *not new principles* that we need but a willingness to accept the consequences of the fact that biological systems occupy a *different region of the space of physical relations than do simpler physico-chemical systems, a region in* which the objects are characterized, first, by a *very great internal physical and chemical heterogeneity* and, second, by a *dynamic*

exchange between processes internal to the objects and the world outside them. That is, organisms are *internally heterogeneous open systems.*"

The interplay of static and dynamic forces to create something as subtle and as intangible as poetry has been phrased in the following way by the Scottish poet George Mackay Brown (1922–1996).[13]

> The minglings of sea and earth
> Creel and plough
> Fish and cornstalk
> Shore people and shepherds
> Are the warp and weft that go
> To make the very stuff of poetry

Genomics, Proteomics, and the Essence of Life: A Faustian Dialog

The publication a few years ago of the entire genome of *Mycobacterium tuberculosis*,[1] the tubercle bacillus responsible for the death of so many members of the human species, represented another milestone in the development of what has been called experimental genomics. In a broad sense, this term refers to the efforts to systematically obtain the complete nucleotide sequence of the genome of important and relevant organisms of the earth ecosystem. Many more complete genomes have been sequenced since *M. tuberculosis*, both unicellular and multicellular, and encompassing pathogens as well as other organisms of medical and economic interest. The ultimate goal is to complete the entire human genome by 2005. Tremendous progress has been achieved on this front and the first draft of the complete human genome has already being deciphered by two separate sequencing efforts.[2,3] The related areas of functional, computational, and structural genomics will complement and enhance the impact of these data gathering efforts (see Chapter 12 on databases).

Only a few years ago, J. Craig Venter published the random sequencing and assembly of the first complete DNA from the free living bacterium *Haemophilus influenzae*, completed by the novel "shotgun cloning."[4] Since then we have seen in rapid succession entire genomes of several microorganisms representing important biological classes ranging from *E. coli*[5] to *B. subtilis*[6] and *S. cerevisiae*.[7] This methodology has recently produced the complete genome sequence of the favorite organism of the classic geneticists *Drosophila melanogaster*[8] and, at the end of the millennium, the first complete genome sequence of a simple flowering plant, the thale cress *Arabidopsis thaliana*.[9]

A glimpse at the figures that keep appearing from the complete projects is sobering. For example, the entire genome of *B. subtilis* (an important member of the class of Gram-positive bacteria) contains 4,214,810 base pairs. They code for approximately 4,100 proteins, 42% of which have unknown functions. Among those encoded proteins there are 18 transcription factors and 77 ATP-binding transport proteins. Many of these new sequences represent important biotechnology findings related to the production and transport of antibiotics.[10]

The amount of information, the quantity and quality of our knowledge of the biological systems that surround us was unimaginable only a few years ago. The idea of obtaining the three-dimensional structure of all (or a sizable subset) of the proteins coded by these millions of bases could whet the appetite of the novice crystallographers and sharpen the tools of the well-seasoned ones (see for example the brief reviews by Chayen & Helliwell[11] and Pennisi).[12] The possibility of mapping structurally the majority of the macromolecular structures playing critical roles in the human body appears to be within reach in the not too distant future. Similarly, obtaining the three-dimensional structure of the entire collection of protein folds (i.e., the universe of protein folds) is driving intense research efforts in what is referred to as "structural genomics".

Undoubtedly, the knowledge so gained will be extremely valuable to find novel antibiotics, to understand the molecular basis of many known and unknown diseases in animals and men, and to design potent and efficacious drugs against them. Detailed knowledge of the genomes of various microorganisms will pave the way for understanding rare metabolic pathways that could help in solving environmental problems. These are just a few of the immediate applications. One can only make conjectures as to the unexpected findings.

All of these futuristic vistas notwithstanding, I can see myself clad in a white robe as a young, Faustian, thirsty-for-insight, macromolecular crystallographer in front of Mephistopheles, having the following conversation:

"I want to know the secret of life."

"I presume you mean beyond the DNA double helix and the genetic code."

"Yes of course. Those are past history."

"What do you want to offer in return?"

"I am willing to sacrifice love or happiness for the ultimate knowledge."

"Do love and happiness not accompany each other?"

"Fine! Let us not go into details. I'll put both on the scale."

"You, macromolecular crystallographers, have your own view of what the secret of life is."

"I suppose we do. I'll be more specific. Can you put at my fingertips, the complete three-dimensional structure of all the macromolecular components encoded in the *E.coli* genome?"

"Of course, I can, but remember your personal love and happiness must be on the other side of the scale. And what is the resolution you wish?"

"Good point. Refined at least to 1.2 Å resolution."

"Certainly, in exchange for your unhappy existence, totally deprived of love. Are you ready? Are you sure you want to do this?"

"Why do you keep on insisting on the conditions? What do you mean am I sure? You are supposed to entice me."

"I mean that if I were to sacrifice my love and my happiness for the secret of life, I'd better be sure of what I really understand by that."

. . . .

Am I sure?

Are we sure?

Albert Szent-Györgyi (1893–1986), the Hungarian biochemist who isolated Vitamin C from extracts of paprika from his native land, wrote in the preface to one of his books the following parable:

> If you give a dynamo to a chemist, the first thing he will do is to submerge it in HCl and analyze what substances are deposited during and after the reaction, and which gases are given away. If you give it to a molecular biologist (of his time), he will take it apart, disassemble it, characterize each and everyone of its parts and then he will put it back together again. Now, he argued, if you were to point out to these scientists that the dynamo works because of changes with time of something called "magnetic flux," they will call you a "vitalist". Namely, a person who needs to invoke an "*élan vital*" to explain biological phenomena.[*]

Any person contemplating a musician playing an instrument can certainly relate to the following analogy. We can take a musical instrument (i.e., an oboe) from the hands of the musician playing it. We can disassemble it and make a detailed analysis of each individual part; then, we can put it back together again. Our language might betray us when we leave the instrument on the table saying that the instrument is "lifeless" without the musician playing it. In fact, it is not the person as a unique demiurge that originates music. Rather, it is the flow of air circulating within the cavities and interstices of the instrument, as channeled and diverted by the action of its valves and keys, that produces the sound that we call music.

Similarly, as important as the molecular components are for life, we should not forget the interplay of flows and forces that support it. The currents of multiple ions, protons, and electrons; the pressures created by chemical gradients; the pulses of electric currents and polarization potentials; and the molecular migrations and myriad of feedback loops. All are critical elements of the living cell. It is pre-

---

[*] I truly believe the book is "*Introduction to a Submolecular Biology*." I apologize to the reader because much to my disappointment, I have not been able to find the complete quote and citation of this comparison. However, what I wrote reflects the essence of the text as it appears in my old notes.

cisely the cessation of these fluxes and rhythms that mark the absence of life. Our constitutive parts will remain long after the termination of the ephemeral wind of life. Our challenge is to understand how our beautiful macromolecular structures permit, facilitate and maintain that fragile and intangible state that we call living.

27

When I Heard the Learn'd Crystallographer

Probably every high school student in the U.S. and in most of the Anglo-Saxon world has been exposed to a poem by the American poet Walt Whitman (Long Island, NY, 1819–1892) entitled *When I Heard the Learn'd Astronomer*. It is a poem that has been included in all too many anthologies and that has been used many times to illustrate the so-called gap between C. P. Snow's two cultures: the sciences and the humanities.

To me however, the poem epitomizes the need for complementary approaches in our perception, depiction and understanding of the cosmos that surrounds us. There is an analytical description of the world based on scientific concepts and ideas, equations and numbers. However, there are also complementary *apprehensions* of the same reality based on various frameworks or conceptualizations. Difficult to describe qualities such as innate beauty, contingency, fragility, and interconnectedness can make certain facets of the natural world more appealing to certain people than its scientific regularities. I would even argue that the same person might enjoy different perceptions of the same phenomenon at different times or even essentially simultaneously. Do we have to know every single note of Handel's *Messiah* to appreciate it? Will all listeners mentally consult the "circle of fifths" while following the modulations of a piano concerto by Mozart? Probably not. Depending upon the time, the occasion and the profession, most of us simply listen to the music and enjoy it.

Similarly, putting aside atoms, molecules, monomers, and polymers, organelles, one can look at the overall cell for further insight, or just for sheer enjoyment. To contemplate cellular structures for what they can be considered to be: molecular masterpieces (**Plate 12.3**), dynamic microcosms of intricate detail (**Plates 14.1** and **21.1**). A protein structure can be described and analyzed in terms of: i) the Cartesian coordinates of the component atoms; ii) the values of all the possible torsion angles present in the structure; iii) the surface accessibility; iv) the charge distribution; v) relative rigidity of the different parts of the structure; and many other ways, including the simple contemplation of its intrinsic chemical wizardry and beauty where each individual atom is perfectly nestled in an environment that perfectly and elegantly complements its properties.

I do not mean to imply that the analytical, reductionistic tools are superfluous. No, and with chemical redundancy, I say $6.023 \times 10^{23*}$ times "No." I am just arguing that there are many different approaches to "comprehend" the molecular underworld beyond our reach, and that the analytical mode is only one of them, although a very important one.

Not coming from an English speaking culture, I discovered the poem rather late, when I was already a professional biophysicist. Nonetheless, it immediately occurred to me that it could be adapted to many other scientific professions. Putting aside the issues of copyright for the moment, I am sure that the bard of Long Island will not object to my modifying his poem for an expansion of its message and its audience. Besides, the reader might have seen the film *Il Postino* (*The Postman*, in the English version, 1995, directed by Michael

---

* 6.023×10^{23} is Avogadro's number. It is the number of molecules of a substance that together will weigh as much as the molecular weight, in grams. It is put in this sentence to emphasize the number of times. This number is essential to chemical calculations.

Radford). In this sensitive romantic story, the main character argues with the famous Chilean poet Pablo Neruda that poetry does not belong to whoever writes it but rather it is the property of the person who needs it.

Therefore, following this dictum, I have taken the liberty of finishing these notes with an adaptation of Whitman's poem to convey the importance of complementary insights and approaches in our perception, analysis, and understanding of the molecular microcosms that surrounds us.

WHEN I HEARD THE LEARN'D CRYSTALLOGRAPHER

When I heard the learn'd crystallographer

When the bonds, the angles, were ranged in columns before me,

When I was shown the charts and diagrams, to add, divide, and measure them,

When I sitting heard the crystallographer where he lectured

with much applause in the lecture-room,

How soon unaccountable I became tired and sick,

Til rising and gliding out I wander'd off by myself,

In the mystical dim-lighted room, and from time to time,

Look'd up in perfect silence at the protein fold.

Adapted from *Leaves of Grass* (Walt Whitman)

May this poem adaptation be my way of condensing and paraphrasing what has been so eloquently expressed by Richard Dawkins in his book *Unweaving the Rainbow: Science, Delusion and the Appetite for Wonder*[1] and to which I alluded before in passing in a previous chapter. Namely, knowing that the constitutive parts of the biosphere are molecules constructed according to the same unavoidable laws of physics and chemistry upon which the entire universe is based, should not prevent us from admiring their intrinsic fabric, texture, and atomic intricacy. An exquisite design and complex architecture, which combined with the appropriate ebbs and flows of other simpler molecules and inanimate ions, transform inert matter into living organisms.

epilogue

Dear Reader, we have reached the end of our journey. What have you learned about crystals? What is then "The Magic of Crystals"? Let us compare notes.

Personally, I think that I have been fascinated with crystals since my childhood but at a subliminal level. I did like the lessons on geology and biology at the high school in my hometown (Aranda de Duero, approximately fifty miles south of the province capital Burgos) with the teacher Ms. Victoria Serrano. From those days, I remember how easy it was for me to remember, understand, and recognize, using a wooden model, the symmetry elements of the holohedral form of the cubic system given by the teacher in a compact, almost inscrutable ($3A^{IV}$, $4A^{III}$, $6A^{II}$, 9-P, 1C), algebraic form. Something that none of my classmates could do. I also remember, almost by heart, a quotation from one of the dense textbooks by D. Salustio Alvarado where he stated that the experiment by Max von Laue demonstrated (at the same time!) the wave nature of the X-rays and the atomic nature of the crystals. Although able to recite the statement by heart, I certainly did not fully understand the meaning or the implications of it.

In retrospect, however, I think that this subliminal love with crystals revealed itself in my first year at the University of Valladolid in Spain. I had chosen sciences as a major and my courses were Mathematics, Physics, Chemistry, Geology, and Biology. Everybody knew that the tough courses were the first three, and that the remaining two were fillers. They were required in case you had to teach some of Geology or Biology in a High School, but nobody will take them seriously. As part of the Geology class we were supposed to build cardboard models of the crystallographic forms (holohedral, hemihedral, and tetartohedral) of the seven crystallographic systems. They came in many sheets of soft white cardboard, neatly folded and enclosed into a white paper envelope. They had to be cut, bent, assembled, and glued together. We were supposed to have an oral exam after Christmas, analyzing any one picked at random by the *"catedrático"* (senior professor). As it happened, I left the residence for the Christmas vacation on the 22 of December by bus. Next day in the utmost despair, I realized that I had forgotten the envelope containing all those precious cutout models of the crystallographic polyhedra in the dormitory, and I was not going to be able to complete the project for the exam. Returning to Valladolid for that purpose alone was out of the question and there was no telephone within reach at the time. What to do? I wrote a polite letter to the director of the residence, explaining to him where the precious envelope was in my room, and I asking him, please, to return it to me by return mail. So he did, and I spent many afternoons in my father's study table assembling and gluing all the models together with my sister Amparo and my younger brother José-Luis. When I realize now that I was seventeen years old at the time, and the number of hours that I spent doing that, I cannot but think that it was love…

All of that being said, I think that to me crystals are magic for two main reasons. First, the symmetry they possess, both internally and externally, makes them beautiful objects. Second, this symmetry gives them unusual properties and allows us to "*see*"—truly see—the atoms and molecules that constitute the micro-cosmos that surrounds us by the interaction of crystals and X-rays. This atomic universe naturally includes ourselves, and the myriad of macromolecules that constitute the living planet, of which we are part. Ironically, by being able to see the atomic intricacies of the molecular constituents of life processes, we can design better drugs and novel vaccines using the atomic information obtained from crystals. In this indirect way, one can say metaphorically that crystals can help to cure diseases and have "healing properties".

After reading all these pages, perhaps crystals have worked their magical spell on you like they did on me many years ago. If so, now is the time to read and study the textbooks, learn the techniques, and use the computer programs. Absorb as much as possible of the known territory and embark on a lifetime adventure of exploration and discovery. You can also participate in this unending dialog with Nature and the study of crystals and crystalline materials is an excellent way to interrogate Her. May the "magic of crystals" always go with you!

glossary

his glossary is only intended to serve as a guide in reading the text. Detailed explanations of the technical terms should be found in the textbooks and references listed in appropriate section.

active site: in a **protein** performing a chemical reaction (**enzyme**), the region on its surface where the chemical reaction takes place. The compound undergoing the reaction (i.e., the **substrate**) binds in this region.

adenine: complex **molecule** of basic character (i.e., a nitrogenous base) that is part of the DNA; it forms hydrogen bonds with **thymine**.

amino acid: the basic repeating unit of a protein **molecule**. Chemically amino acids are composed of C, O, N, H, and S. They contain an acid group (–COOH), an amine group (–NH$_2$), and a central carbon (C$_\alpha$) containing various substituents (–R) that give each amino acid a set of unique chemical properties. Diagram of the chemical structure of an amino acid:

Amino acid: side chain R$_1$
Planar representation

Amino acid: side chain R$_1$
3-D representation: L

Amino acid: side chain R$_1$
3-D representation: D

Individual amino-acids have different R groups. Examples:

  Glycine: –H; Serine: –CH$_2$OH;
  Cysteine: –CH2SH; Aspartic: –CH$_2$–COOH;
  Alanine: –CH$_3$.

For a three-dimensional representation of several amino acids using computer graphics see **Plates 7.1, 11.1, 20.1, 20.2**, and **21.1**.

asymmetric unit: the smallest region of space that can generate a **crystal**. Atomic or molecular entities are placed in or occupy the asymmetric unit.

ATP: abbreviation for adenosine tri-phosphate. A complex chemical entity that acts as an energy carrier in all living organisms. Conceptually an ATP **molecule** contains three parts: an **adenine** (a complex base), a sugar containing five carbon atoms (ribose), and a string of three phosphate groups connected together. Controlled cleavage or addition of these phosphates is used in biochemical reactions to release or store energy. A similar molecule is **GTP** where only the first component, the base (i.e., **guanine**), is different. GTP is commonly involved in the transfer of chemical signals in the cell.

Avogadro's number N_A: the number of **molecules** in a mol (gram molecule) of substance. $N_A = 6.023 \times 10^{23}$. This number is essential to chemical calculations.

Bragg's law: the cornerstone of crystallography. Reduces the analysis of diffraction patterns to one geometric variable d_{hkl}—the spacing of a set of parallel planes with indices h,k,l—and one physical parameter from the X-ray source, λ—the wavelength. This physical law states that if two parallel **X-rays** of wavelength λ are reflected by adjacent planes, distance d apart in a crystal lattice, the X-rays will constructively interfere when

$$2\,d_{hkl} \sin \theta = n\,\lambda,$$

where n is an integer and θ is the angle (called Bragg's angle) that refers to the direction of incidence of the X-rays upon the surface. If n is fractional, then destructive interference is observed.

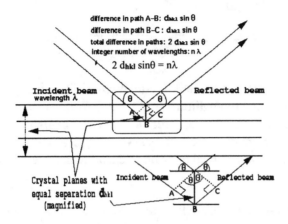

difference in path A–B: d_{hkl} sin θ
difference in path B–C: d_{hkl} sin θ
total difference in paths: 2 d_{hkl} sin θ
integer number of wavelengths: $n\lambda$

$$2\, d_{hkl}\, \sin\theta = n\lambda$$

Incident beam wavelength λ

Reflected beam

Crystal planes with equal separation d_{hkl} (magnified)

Incident beam

Reflected beam

The law is named Bragg's law after its best-known discoverer W. Lawrence Bragg (1890–1971), the son of the famous British lographer Sir William Henry Bragg (1862–1942).

capillary tube: a very thin, hollow, tube made of glass or quartz.

co-enzymes: see **co-factors**.

co-factors or co-enzymes: small (in comparison to **proteins** that contain thousands of atoms) organic **molecules** that act in conjunction with proteins to catalyze the innumerable reactions taking place in living organisms. Many vitamins are co-factors that need to be ingested in the diet because the body cannot synthesize them.

collimator: narrow tube used to direct the **X-rays** towards the sample (typically a crystal or crystalline sample).

coordinates: set of three-ordered numbers (x,y,z) that are used to place a point in three-dimensional space.

cos *x*: a mathematical function describing a simple and uniform wave; cos *nx*, (*n* an integer number) represents a similar wave with a different frequency of oscillation (higher pitch for numbers larger than one). (see **sin *x***).

crystal(s): a regularly shaped object with internal three-dimensional **symmetry**.

crystallographic refinement (refinement): iterative process by which the position (**coordinates** *x,y,z*) and dynamic parameters (**temperature factors**) of the constitutive atoms of a **protein** are best fitted to the data obtained from the **crystals**. Typically requires cycles of computation and manual revision of the protein models to optimize the fit to the data.

cubic or isometric: crystallographic system of highest **symmetry** characterized by two series of conditions—all the angles are orthogonal and the three edges of the crystal lattice are equal (a = b = c). This translates into the unique feature of the cubic system that is the presence of four intersecting three-fold symmetry axes.

cytosine: complex base that forms part of the DNA **molecule**. It typically pairs with **guanine**.

C_α carbon: the central carbon of one **amino acid**, the basic building block of protein **molecules**. From this central carbon four different chemical groups or substituents extend out. If this central carbon occupies the center of a cube, the four branches will extend towards four non-consecutive corners of a cube forming a tetrahedron. In other words, if the central carbon is placed at the center of a tetrahedron, the four branches will extend to the corners, just like the arms of a child's jack. For a three-dimensional representation of several amino acids using computer graphics see **Plates 11.1, 20.1, 20.2,** and **21.1.** The spatial arrangement of four different groups on a tetrahedron results on the existence of two different isomers: L- and D-stereo isomers (amino acid).

diffraction pattern: the pattern of regularly-spaced spots that result when **crystals** are exposed to intense **X-rays**.

EC$_{50}$: effective concentration to achieve 50% inhibition in the assay being performed. This parameter is used to screen for active **molecules** in assays using cells.

electron density map (map): a three-dimensional representation of the number of electrons per cubic Å (electrons/Å^3) in the **crystal**. In a two-dimensional representation, electron density maps are like topographical maps, with the regions of high electron density maps corresponding to the mountains, and the low regions the valleys. Regions of high electron density correspond to the position of the atoms. In three dimensions, the image of the polypeptide chain in terms of its electron density is like a thick ribbon with decorations extending out from it at medium resolution.

enzyme(s): a protein **molecule** that facilitates some chemical reaction as a catalyst; it is not consumed in the reaction but recycled back to its original form after the reaction is over.

GTP: abbreviation for guanosine tri-phosphate. A chemical entity similar to **ATP** that acts as a carrier of chemical signals in more complex cells. Structurally GTP differs from ATP only in the character of the complex base—**guanine** as opposed to **adenine**. Controlled cleavage of the terminal phosphate group is used to transfer chemical signals within the cell.

guanine: base component of DNA. It pairs with **cytosine** in the DNA helix.

hemihedral: a crystallographic form that contains only half of the symmetry elements of its class.

hexagonal: crystallographic system characterized by a principal axis of **symmetry** that repeats a **motif** every 60°: 6-fold symmetry. (a = b; not necessarily equal to c).

holohedral: a crystallographic form that contains all the **symmetry elements** of its class.

***h,k,l*:** see **Miller indices**.

IC$_{50}$: inhibitory concentration that achieves 50% inhibition of the **enzyme** being assayed, typically in an *in vitro* assay.

icosahedral asymmetric unit: by extension from the crystallographic asymmetric unit, defined earlier, it is the unique portion that allows the construction of a solid containing icosahedral symmetry by application of the corresponding **symmetry operations**.

isometric: see **cubic**.

K$_i$: inhibition constant of a compound in an enzymatic assay, under very concrete assay conditions.

ligases: enzymes that permit the joining of fragments of DNA.

macromolecule(s): very large **molecule** made up of thousands of atoms with an internal monomeric building-block repeated many times, giving rise to **polymers**. The different types of plastic are polymers of small distinct monomers.

MAD: see **multiwavelength anomalous diffraction**.

map: see **electron density map**.

median: in an ordered set of values or in an ordered set of frequency intervals, the value in the middle. There are as many values above it as there are below it. For example, in the ordered set of numbers 1,2,3,4,5; 3 is the median (two above and two below it).

methionine: a sulfur-containing **amino acid** that can be replaced in a **protein** by a selenium-containing one **(Se-Methionine)**, using techniques of **molecular biology**.

This atomic replacement is critical for the fast determination of protein structures by the method of **Multiwavelength Anomalous Diffraction** (**MAD**) using tunable sources of **X-rays** such as synchrotrons.

MIC$_{90}$: minimum inhibitory concentration to achieve 90% inhibition of the **enzyme** being assayed.

Miller indices: the set of three integers *h,k,l* used by crystallographers to characterize a plane or set of parallel planes, or face of a **crystal**. This set of three indices labels each and every one of the diffraction spots present on a **diffraction pattern**.

MIR: see **multiple isomorphous replacement**.

mode: in a distribution of numbers or distribution of frequency bins (intervals), the value that occurs most often.

molecule(s): group of atoms held together in a stable form.

molecular biology: in general terms, refers to the part of biology that attempts to explain the functioning of living beings in terms of their constitutive **molecules**. However, the same term is used routinely to express the methodology that uses and manipulates DNA molecules or fragments to produce new genes.

molecular genetics: refers to the synthesis, regulation, operation, and manipulation of DNA.

molecular replacement (MR): computational method to obtain the three-dimensional structure of **macromolecules** proposed and developed by Michael G. Rossmann. It makes possible mapping the three-dimensional structure of unknown macromolecules, using the knowledge of structurally-related **molecules**. The method uses only the data from the native **crystals** containing the three **Miller indices** (*h,k,l*) and the structure factor (**F$_0$**) derived from the intensity of the reflection with those three spatial **coordinates**. This method uses computational techniques to bypass the "**phase problem**."

monoclinic: the name from the Greek, *mono-klinos*, means one inclined angle. Crystallographic system characterized by a principal axis of **symmetry** that repeats a **motif** every 180°—2-fold symmetry: an axis not necessarily equal to b nor c. A good example of a monoclinic cell is a box used to pack to two light bulbs together. The sides of the cardboard box are rectangular (90°) but the box itself can be squeezed to almost any angle, including 90°.

motif: a recurring, dominant element. In **crystals**, the motifs are atoms, clusters of atoms or **molecules**, which generate the crystal by the appropriate **symmetry operations**.

MR: see **molecular replacement**.

multiwavelength anomalous diffraction (MAD): experimental method of obtaining the three-dimensional structure of **macromolecules**. It is based on obtaining diffraction data from **crystals** containing an anomalous X-ray scatterer. Typically three data sets are collected around the X-Ray absorption edge of the atom scattering anomalously. The use of selenium as a preferred anomalous scatterer was proposed by Wayne A. Hendrickson and it is the method of choice for a rapid determination of protein structures, if crystals containing selenium can be obtained.

multiple isomorphous replacement (MIR): experimental method of obtaining the three-dimensional structure of **macromolecules**. It is based on the introduction of several heavy (strong scatterers of **X-rays**) atoms into the crystal lattice. The changes in the measurements of the diffraction data are used to compute phases of the individual reflections. The values of the diffracted intensities and their estimated phases are used to reconstruct the electron density in the **crystals**. This method was pioneered by Max Perutz and John Kendrew and their co-workers in the structure determinations of hemoglobin and myoglobin.

non-crystallographic symmetry: **symmetry** exhibited by a group of **molecules**—typically **proteins**—that is not generated by the **symmetry operations** of the **crystal**. The additional symmetry can be used to accelerate the determination of the three-dimensional structure of complex molecular aggregates (i.e. viruses) by averaging procedures. These concepts are related to the ideas of the **molecular replacement** method.

orthorhombic: the name from the Greek *ortho-rhombus*, is referring to a rhombus with three orthogonal (90°) angles. Crystallographic system characterized by three ninety-degree angles among the axes without limitations on the dimensions. The axis a not necessarily equal to b nor c. In other words, shaped like a brick or shoebox.

peptide bond: chemical bond that connects together **amino acids** to form a protein chain. The order of amino acids in a protein chain is always described by convention by starting at the amino acid with a free amine ($-NH_2$) group. Chemically, the formation of a peptide bond is the coupling of two amino acids together by the removal of one **molecule** of water between the acid group ($-COOH$) of one amino acid and the amine group ($-NH_2$) of the next in the chain:

Peptide bond

NH₂——C==COOH +NH₂——C——COOH $\xrightarrow[+H_2O]{-H_2O}$ NH₂——C——C—N——C——COOH

| Amino acid 1 | Amino acid 2 | Dipeptide: two amino acids linked together |

Peptide bonds ($-C(=O)-N(-H)$) and fragments of a polypeptide chain are illustrated, using computer graphics in **Plates 7.2, 11.1, 20.1, 20.2, and 21.1.**

phase problem: the crucial problem faced by crystallographers. It derives from the fact that the diffraction experiments can only measure the intensity of the diffraction peaks and not their relative phases.

plasmid: a circular fragment of DNA that genetic engineers use to insert foreign pieces of DNA into bacteria.

polymer(s): a chemical entity formed by the repetition of a basic unit or monomer, **proteins** are polymers. Many plastics are polymers of a basic chemical unit.

protein(s): a protein is a **polymer** composed of individual monomers (**amino acids**) joined in a linear chain. Each protein has a unique amino acid sequence and a corresponding three-dimensional structure. Proteins are essential components of living beings, composed of many (hundreds to thousands) atoms. Hair is made up of protein; milk, eggs, and meat all contain proteins of various types and properties.

protein crystal: a **crystal** made up of protein **molecules**.

refinement: see **crystallographic refinement**.

real lattice (real space): the ordered array of points that contain the atomic structures of **molecules**. Within this array of points, set of planes are characterized by the common direction, given by three integer numbers h,k,l (**Miller indices**) and the constant interplanar distance (d_{hkl}).

real space: see **real lattice**.

reciprocal lattice (reciprocal space): the ordered array of points derived from the direct lattice by defining a reciprocal lattice point for each set of parallel equidistant planes. The common direction is taken normal to the planes and is defined by three integer numbers (h,k,l); in the reciprocal lattice, the interplanar spacing is taken as proportional to the inverse (reciprocal) of the direct distance space spacing (i.e., K/d_{hkl}, with K being an appropriate constant).

reciprocal space: see **reciprocal lattice.**

restriction enzymes: **enzymes** that permit the cutting of DNA at specific regions distinguished by their local sequences of nucleic acid bases. For example, the **enzyme** *Bam*HI isolated from *Bacillus amyloliquefaciens* H (enzyme I) cuts the DNA at the sequences:

G^GATCC   (5'->3')

CCTAG^G   (3'->5')

as indicated, leaving uneven ends. This permits the insertion of other fragments of DNA into the DNA **plasmids.** The symbol ^ indicates the position of the cut in the DNA chain.

ribosome: the crucial component in the machinery of the cell that synthesizes **proteins.** It is a very large macromolecular complex containing proteins and nucleic acids of a special type named ribosomal RNA (rRNA). DNA and RNA are other nucleic acids. The entire ribosome of simple bacteria is composed of two parts or subunits referred to as 50S and 30S to make the entire ribosome.

sin *x*: a mathematical function describing a regular wave; sin *nx* (*n* is an integer number), represents a similar wave with a faster frequency of oscillation (higher pitch). Waves described by cos *nx* and sin *nx* can be combined with appropriate coefficients to produce almost any repeating shape. For instance, you can try on a graphing calculator: $6 \cdot \cos 3x + 8 \cdot \sin 5x$; 6 and 8 are the coefficients of this simple "linear combination" of two waves. The factors 6 and 8 are numerical values related to the relative height (intensity) of the waves in the combination. 3 and 5 specify different oscillation rates for the two waves. (See also **cos *x*.**)

storage ring: see **synchrotron or storage ring.**

substrate: chemical compound(s) undergoing a chemical reaction. Typically one or more substrates (A, B) react to yield one or more product(s) (P). In biological systems, chemical reactions are facilitated by **enzymes**, which are **proteins** acting as catalysts.

$$A + B \xrightarrow{E} P$$

The reaction could happen without the **enzyme**, but very much more slowly.

symmetry: it means repetition as found in wall paper designs, fabric patterns, embroidery and graphic art and buildings. There are many kinds of symmetry: many different ways of repeating one idea or **motif**. Wall paper designs and fabric patterns are created by translational symmetry in two-dimensions. **Crystals** are characterized by translational symmetry in three directions. In buildings, many organisms, and humans, the symmetry is bi-lateral: one half is identical to the other half, by a mirror reflection across a central plane.

Rotational symmetry is found in steering wheels, plates, bowls, and dinnerware because these objects are symmetrical around a rotation axis. Patterns with motifs repeated every sixty degrees (**hexagonal** symmetry) are very common.

symmetry operation(s): the process by which an image or **motif** is duplicated without distortion. There are many types of symmetry operations. Common operations are rotation and displacement or translation. Symmetry operations are easy to visualize in a plane (i.e., on a tabletop) but in general they take place in three dimensions. Take an irregular object, for example a pair of scissors, and rotate it around its center point by 60°, 90°, and 120°; those are rotational symmetry operations. Instead, just move it from one place to another maintaining the orientation and this will be a symmetry operation consisting of translation only. Of course you can combine them.

synchrotron or storage ring: a particle accelerator where the electrons are kept circulating around a nearly circular orbit; while doing so, electrons give off **X-rays** in a direction tangential to the orbit of the electrons. Experimental stations are built around a synchrotron along the path of the X-rays given off by the electrons moving around in the storage ring. Each individual station is designed with special characteristics to utilize the properties of the X-rays in different ways.

temperature factor: a fourth parameter (in addition to the **coordinates** x,y,z) used by crystallographers to characterize the motions of an atom near its equilibrium position.

tetartohedral: a crystallographic form that contains only one-fourth of the symmetry elements of its class.

tetragonal: crystallographic system characterized by a principal axis of **symmetry** that repeats a **motif** every 90°—4-fold symmetry (a = b; not necessarily equal to c).

thymine: base in the DNA structure. It pairs with **adenine**.

trigonal: crystallographic system characterized by a principal axis of **symmetry** that repeats a **motif** every 120°—3-fold symmetry. (a = b; not necessarily equal to c).

triclinic: simplest crystallographic system characterized only by the repetition of a **motif** in three dimensions. The name (from the Greek, *tri-klinos*) means three inclined (or three non-orthogonal—different from 90°) angles. The three edges of the crystal cell (a, b, and c) can have different values.

unit cell: the box that contains the basic building block of a **crystal**. The crystal is generated by translation of this block along the three axes.

X-rays: a very penetrating form of radiation, as used in hospitals. The difference between X-rays and other forms of electromagnetic radiation (sun rays) is the wavelength (λ) or frequency (ν : $\lambda = c / \nu$; c is the speed of light in vacuum: 300,000 Km/sec). The different colors of the rainbow that combine to make up white light have different frequencies going from low (reds) to high (blues). Invisible frequencies to the human eye extend below the red (infrared) and beyond the violet (ultraviolet). The energy of a radiation is given by Planck's equation $E = h\nu$ ($h = 6.62\ 10^{-27}$ erg sec, is the Plank's constant; ν is the frequency of the radiation; (Max K. E. Planck, the German physicist, 1858–1947). Red is the "low energy" portion of the visible spectrum while violet is the "high energy" region. X-rays have a wavelength shorter (i.e., larger frequency) than ultraviolet radiation by approximately hundred-fold.

X-ray detector: an instrument or device that detects or records X-rays. The photographic plates used at the hospitals are a type of detector. Modern detectors use electronic circuitry.

X-ray generator: a scientific instrument that produces X-rays.

α-helix: one of the basic folding patterns of the polypeptide chain characterized by a helical coiling, much like an old-fashioned telephone cord. This structure is held together by hydrogen bonds formed by **amino acids** along the helical axis of the same chain.

β-sheet: the second of the basic folding patterns of the polypeptide chain, characterized by a wavy pattern, much like a pleated sheet. This structure is held together by hydrogen bonds formed by **amino acids** from different chains, approximately perpendicular to the length of the polypeptide chains.

bibliography

References

preface

1. *Biological Sciences.* (Good Readings series). New American Library: New York, 1956.

   Reference found in the introduction of the book:

   Lost Woods. The Discovered Writings of Rachel Carson. Edited and with introductions by Linda Lear. Beacon Press: Boston. 1998, pp. xii–xiii.

chapter 1

1. Symes R.F., Harding R.R. *Crystal and Gem.* (Eyewitness books No. 25.) Alfred A Knopf, Inc. New York. 1991.

   Originally published in the United Kingdom by Dorling Kindersley Ltd Publishing, London.

2. Michele V. de. *Crystals: Symmetry in the Mineral Kingdom.* Crescent Books: New York, 1972.

   Original book printed by IGDA, Novara, 1967. The translation copyrighted in 1972 by Crown Publishers Inc.

3. http://www.lmcp.jussieu.fr

   Web site of the Laboratory of Mineralogy and Crystallography of Paris, France, where an excellent mineral and crystal collection is exhibited. This site contains lots of valuable information about the early history of crystallography as published in a beautifully illustrated booklet entitled *200 Ans de Crystallographie en France* published on the occasion of the XVth Congress of IUCr that took place in Bordeaux, France, in July 1990.

4. Sobel, D. **Longitude. The True Story of a Lone Genius Who Solved the Greatest Scientific Problem of His Time.** Walker and Company: New York, 1995.

5. *The International Tables for Crystallography.* Vol. A. *Space-Group Symmetry.* Hahn, Theo, ed., 5th edition. IUCr/Kluwer Academic Publishers: Dordrecht-Boston-London, 2002.

6. *The International Tables for Crystallography. Brief Teaching Edition of Volume A.* Hahn, Theo, ed., 5th revised edition. IUCr/Kluwer Academic Publishers: Dordrecht-Boston-London, 2002.

chapter 2

1. Seliger H.H. **Wilhem Conrad Roentgen and the Glimmer of Light.** *Physics Today* 1995, Special issue: *X Rays 100 Years Later* (November 1995), p. 25.

2. Heilbron J.L., Bynum W.F. **Eighteen ninety five and all that.** *Nature* 1995, **373**:p. 11–12.

3. See for instance: Glusker J.P. and Trueblood K.N. *Crystal Structure Analysis. A Primer.* 2nd edition. Oxford University Press: New York–Oxford, 1985, Legend to Fig. 2.3 on p. 17.

chapter 3

1. *The Norton/Grove Concise Encyclopedia of Music.* Stanley Sadie, ed., Alison Latham, assistant ed., W. W. Norton and Co.: New York–London, 1988.

2. ASRL Web site (http://www.artscienceresearchlab.org).

3. *The World of M. C. Escher.* New Concise New American Library edition. With texts by M. C. Escher and J. L. Locher. Harry N. Abrams, Inc. Publishers: New York, 1974. p. 141.

4. Ibid., p. 143.

5. MacGillavry C.H. *Fantasy and Symmetry. The Periodic Drawings of M. C. Escher.* Harry N. Abrams, Inc. Publishers: New York, 1976, p. 62, Plate no. 29.

   Copyright 1965 and 1976 by the International Union of Crystallography, Geneva, Switzerland.

6. Gould S.J. *Full House. The Spread of Excellence from Plato to Darwin.* Three Rivers Press: New York. 1996, pp. 55–56.

chapter 4

1. Herivel J. *Joseph Fourier, The Man and the Physicist.* Clarendon Press: Oxford, 1975.

2. Perutz M. *Protein Structure. New Approaches to Disease and Therapy.* W. H. Freeman and Co.: New York, 1992, Appendix 1. pp. 261–275, Fig. A1.1.

3. Rossmann M.G. **From the Structures of Simple Salts to Those of Sophisticated Viruses.** In: *Crystallography Across the Sciences.* H. Schenk, ed. Munksgaard Intenational: Copenhagen, 1998, p. 717.

chapter 5

1. Atkins P.W. *The Periodic Kingdom. A Journey into the Land of the Chemical Elements.* Basic Books, A Division of Harper Collins Publishers: New York, 1995.

2. Hoffmann R. *The Same and not the Same.* Columbia University Press: New York, 1995.

3. Salem L. *Marvels of the Molecule.* VCH Publishers, Inc. New York, 1987.

   Translation of L. Salem. *Molécule, la merveilleuse.*

   A delightful little book about the internal structure of atoms and molecules in very simple and descriptive language. Distributed worldwide by VCH Verlagsgesellschaft GmbH P.O. Box 1260/1280, D-6940 Weinheim, Germany.

4. Bernal J.D., Crowfoot D.C. **X-ray photograph of crystalline pepsin.** *Nature* 1934, **794**:133–134.

5. Blundell T.L., Johnson L.N. *Protein Crystallography.* Academic Press: New York–London–San Francisco, 1976.

6. Garman E., Lee S. **Cover illustration.** *J Appl Cryst* 1997, **30**:Cover.

7. Bussiere D., Muchmore S., Dealwis C.G., et al. **Crystal structure of ErmC', an rRNA methyltransferase which mediates antibiotic resistance in bacteria.** *Biochemistry* 1998, **37**:7103–7112.

chapter 6

1. Weyl H. *Symmetry.* Princeton University Press: Princeton, New Jersey, 1952.

2. Hargittai I., Hargittai M. *Symmetry. A Unifying Concept.* Shelter Publications, Inc.:Bolinas, California, 1994.

3. MacGillavry, C.H. *Fantasy and Symmetry. The Periodic Drawings of M. C. Escher.* Harry N. Abrams, Inc. Publishers: New York, 1976.

4. Escher M.C. et al. *Escher on Escher. Exploring the Infinite.* Translated from the Dutch by Karin Ford. Harry N. Abrams, Inc. Publishers: New York, 1989.

5. Ibid., p. 83.

6. Ibid., p. 25.

7. Bronowski J. *The Ascent of Man.* Little, Brown and C°: Boston, Toronto, 1973.

chapter 7

1. *The Quotable Einstein.* Collected and Edited by A. Calaprice. Princeton University Press: Princeton, New Jersey, 1996, p. 169.

2. Waal F.B.M. de. **Cultural primatology comes of age.** *Nature* 1999, **399**:635–636; Whiten A., Goodall J., McGrew W.C. et al. **Cultures in chimpanzees.** *Nature* 1999, **399**:682–685.

3. Jahme C. *Beauty and the Beasts.* Virago Press: London, 2000.

4. Jahme C. **Jungle fever.** *The Manchester Guardian Weekly,* Jun 6–14, 2000, p. 21.

5. Lindsey J. et al. *Jane Goodall. 40 Years at Gombe: A Tribute to Four Decades of Wildlife Research, Education and Conservation.* Produced in association with the Jane Goodall Institute. Stewart, Tabori and Chang: New York, 1999.

6. Podjarny A. Personal communication and presentation at the 2000 Annual Meeting of the ACA, St. Paul Minnesota.

chapter 8

1. Nahin, P.J. *An Imaginary Tale. The Story of Square Root of –1.* Princeton University Press: Princeton New Jersey, 1998.

2. Woolfson M.M. *Direct Methods in Crystallography.* Oxford University Press: New York, 1961.

3. Judson H.F. The *Eighth day of Creation. Makers of the Revolution in Biology.* Jonathan Cape Ltd.: London, 1979.

4. Perutz M. *Protein Structure. New Approaches to Disease and Therapy.* W.H. Freeman and Co.: New York, 1992.

5. Ibid. p. 16.

chapter 9

1. Kurlansky, M. *The Basque History of the World.* Walker and Company: New York. 1999.

chapter 10

1. *The Molecular Replacement Method.* M.G. Rossmann, ed. International Science Review Series. Gordon and Breach: New York, 1972, p. 5.

2. Ibid. p. 14.

3. Drenth J. *Principles of Protein X-ray Crystallography.* Springer Verlag: New York-Berlin. 1999.

4. The current Protein Data Bank site (http://www.rcsb.org).

5. Abad-Zapatero C., Griffith J.P., Sussman J.L., Rossmann M.G. **Refined crystal structure of dogfish M4 apo-lactate dehydrogenase.** *J Mol Biol* 1987, **198**:445–467.

chapter 11

1. Tolkien J.R.R. *The Hobbit.* 2nd Edition. Illustrated by M. Hague. Houghton Mifflin Company: New York, 1994.

2. Tolkien J.R.R. *The Lord of the Rings. Trilogy.* **Part Three: The Return of the King.** Houghton Mifflin Co.: Boston, Massachussetts, 2001, Appendix F.

3. Tolkien J.R.R. *The Hobbit.* ibidem p. 2.

4. Neimark A.E. *Myth Maker: J. R. R. Tolkien.* Harcourt Brace & Co: New York, 1996, pp. 85–86.

5. Tolkien J.R.R. *The Lord of the Rings. Trilogy.* **Part One: The Fellowship of the Ring.** Houghton Mifflin Co.: Boston, Massachussetts, 2001.

   Authorized edition of the fantasy classic by Ballantine Books, 1993. Introduction by Peter Beagle. Foreword by J.R.R. Tolkien.

6. Jones T.A. **Interactive computer graphics: FRODO.** *Meth Enzymol* 1985, **115**:157–171.

7. Richards F.M. **Optical matching of physical models and electron density maps: Early developments.** *Meth Enzymol* 1985, **115**:145–154.

8. Salemme F.R. **Some minor refinements on the Richards optical comparator and methods for model coordinate measurement.** *Meth Enzymol* 1985, **115**:154–157.

9. Cowburn D., Riddihough G. **A never ending frontier.** *Nat Struct Biol* 1997, **4**:761–762.

10. Sigler P.B. **A road not taken.** *Nat Struct Biol* 1998, **5**:266.

11. Jones T.A. **A graphics model building and refinement system for macromolecules.** *J Appl Cryst* 1978, **11**:268.

chapter 12

1. Crystallography, Protein Data Bank (announcement). *Nat New Biol* 1971, **233**:223.

2. Bairoch A. Serendipity in bioinformatics, the tribulations of a Swiss bioinformatician through exciting times! *Bioinformatics* 2000, **16**(1):48–64.

3. The *Arabidopsis* Genome Initiative. **Analysis of the genome sequence of the flowering plant *Arabidopsis thaliana.*** *Nature* 2000, 408:796–815.

4. Adams M.D. The **Genome sequence of *Drosophila melagonaster.*** *Science* 2000, **287**:2185–2195.

5. Mewes H.W. et al. **Overview of the yeast genome.** *Nature* 1997, **387** Suppl:7–65.

6. The *C. elegans* Sequencing Consortium. **Sequence analysis of the genome of *C. elegans.*** Science 1998, **282**:2012–2018.

7. Heidelberg J. F. et al. **DNA sequence of both chromosomes of the cholera pathogen *Vibrio cholerae.*** *Nature* 2000, **406**:477–483.

8. Fleischmann R.D. et al. **Whole-genome random sequencing and assembly of *Haemophilus inlfuenzae* Rd.** *Science* 1995, **269**:496–512.

9. International Genome Sequencing Consortium. **The Human Genome. Initial sequencing and analysis of the human genome.** *Nature* 2001, **409**:860–921.

10. Venter J.C. et al. **The Human Genome.** *Science* 2001, **291**:1304–1351.

11. Smith T. **A new era.** *Nat Struct Biol* 2000, *Structural Genomics* **7** Suppl:927.

12. Goodsell D.S. *Our Molecular Nature. The Body's Motors, Machines and Messages.* Copernicus. An Imprint of Springer-Verlag: New York, 1996.

13. *IUCr Newsletter* 2001, **9**(1):26.
    For a complete listing of the PDB entries in this image.

chapter 13

1. Perutz M.F. **The haemoglobin molecule.** Sci American 1964, **211**:64–76.

2. Abad-Zapatero C., Lin C.T. **Statistical descriptors for the size and shape of globular proteins.** *Biopolymers* 1990, **29**:1745–1754.

3. Lewis M., Rees D.C. **Fractal Structures of proteins.** *Science* 1985, **230**:1163–1165.

4. Bailey M., San Diego Supercomputer Center (e-mail: mjb@sdsc.edu).

5. Abad-Zapatero C. et al. **Structure of a secreted aspartic protease from *C. albicans* complexed with a potent inhibitor: Implications for the design of antifungal agents.** *Prot Sci* 1996, **5**:640–652. PDB code **1zap**.

chapter 14

1. Branden C., Tooze J. *Introduction to Protein Structure.* Garlan Publishing, Inc.: New York–London, 1991, p. 204 See p. 235 in the 1999 edition.

2. Ackerman D. *The Natural History of the Senses.* Vintage Books. A division of Random House, Inc.: New York. 1990.

3. Shepherd G.M. *Neurobiology.* 2nd edition. Oxford University Press: Oxford, UK, 1988.

4. Pierret Perret. **Tarn-et-Garonne.** *Atlas Magazine*, Air France, Sept. 20, 1995.

5. Palczewski K. et al. **Crystal structure of rhodopsin: A G Protein-coupled receptor.** *Science* 2000, **289**:739–745.

6. Doyle D.A. et al. **The Structure of the potassium channel: Molecular basis of K^+ conduction and selectivity.** *Science* 1998, **280**:69–77.

7. Tsukihara T. et al. **The whole structure of the 13-subunit oxidized cytochrome *c* oxidaze at 2.8 Å.** *Science* 1996, **272**:1136–1144.

8. Abrahams J.P., Leslie A.G.W., Lutter R., Walker J.E. **ATP synthetase.** *Nature* 1994, **370**:621.

9. Stock D., Leslie A.G.W., Walker J.E. **Molecular architecture of the rotary motor in ATP synthase.** *Science* 1999, **286**:1700–1705.

10. Cowan S.W. et al. **Crystal structures explain functional properties of two *E. coli* porins.** *Nature* 1992, **358**:727–733.

11. Ashcroft F.M. *Ion Channels and Disease: Channelopathies.* Academic Press: San Diego, 2001.

12. Girvin M.E., Rastogi V.K., Abildgaard J.L., Markley J.L., Fillingame R.H. **Solution structure of the transmembrane H^+—translocating subunit *c* of the F1F0 synthase.** *Biochemistry* 1998, **37**:8817.

13. Deisenhofer J., Epp O., Miki K., Huber R., Michel H. **Structure of the protein subunits in the photosynthetic reaction centre of *Rhodopseudonomas viridis* at 3 Å resolution.** *Nature* 1985, **318**:618–624.

14. Palczewski K., Kumasaka T., Hori T., Behnke C.A., Motoshima H., Fox B.A., Trong I.L, Teller D.C., Okada T., Stenkamp R.E., Yamamoto M., Miyano M. **Crystal structure of rhodopsin: A G protein-coupled receptor.** *Science* **289**:739–745.

15. Cowan S.W., Schirmer T., Rummel G., Steiert M., Ghosh R., Pauptit R.A., Jansonious J.N., Rosenbusch J.P. **Crystal structures explain functional properties of two *E. coli* porins.** *Nature* 1992, **358**:727–733.

16. Doyle D.A., Morais Cabral J., Pfuetzner R.A., Kuo A., Gulbis J.M., Cohen S.L., Chait B.T., Mackinnon R. **The structure of the potassium channel molecular basis of K^+ conduction and selectivity.** *Science* 1998, **280**:69–77.

chapter 15

1. Helliwell J.R. *Macromolecular Crystallography with Synchrotron Radiation.* Cambridge University Press: Cambridge, Cambridge, UK, 1992.

2. Hendrickson W. **X rays in molecular biophysics.** *Physics Today* 1995, **48**:42–48.

3. Muchmore S. et al. **X-ray and NMR structure of human Bcl-XL, an inhibitor of programmed cell death.** *Nature* 1996, **381**:335–341.

4. Tsukihara T. et al. **The whole structure of the 13-subunit oxidized cytochrome *c* oxidaze at 2.8 Å.** *Science* 1996, **272**:1136–1144.

5. Scott W.G., Finch J.T., Klug A. **Three-dimensional structure of an all-RNA Hammerhead ribozyme: A proposed mechanism for RNA catalytic cleavage.** *Cell* 1995, **81**:991–1002.

6. Cate J.H., Gooding A.R., Podell E., Zhou K., Golden B.L., Kundrot C.E., Cech T.R., Doudna J.A. **Crystal structure of a group I ribozyme domain: Principles of RNA packing.** *Science* 1996, **273**:1678–1685.

7. Geiger J.H., Hahn S., Lee S., Sigler P.B. **Crystal structure of the yeast TFIIA/TBP/DNA complex.** *Science* 1996, **272**:830–836.

8. Hunt J.F., Weaver A.J., Landy S.J., Gierasch L., Deisenhofer J. **The crystal structure of the GroES co-chaperonin at 2.8 Å resolution.** *Nature* 1996, **379**:37–45.

9. Pennisi E. **News focus: The race to the ribosome structure.** *Science* 1999, **285**:2048–2051.

10. Garret R. **Mechanics of the ribosome.** *Nature* 1999, **400**:811–812.

11. Clemons Jr. W.M et al. **Structural of a bacterial 30S ribosomal subunit at 5.5 Å resolution.** *Nature* 1999, **400**:833–840.

12. Ban N. et al. **Placement of protein and RNA structures into a 5 Å-resolution map of the 50S ribosomal subunit.** *Nature* 1999, **400**:841–847.

13. Cate J.H. et al. **X-ray crystal structures of 70S ribosomal complexes.** *Science* 1999, **285**:2095–2104.

14. Ban N., Nissen P., Hansen J., Moore P.B., Steitz T.A. **The complete atomic structure of the large ribosomal subunit at 2.4 Å resolution.** *Science* 2000, **289**: 905–920.

15. Wimberly B.T., Brodersen D.E., Clemons W.M., Morgan-Warren R.J., Carter A.P., Vonrhein C., Hartsch T., Ramakrishnan V. **Structure of the 30S ribosomal subunit.** *Nature* 2000, **407**:327–339.

16. Schluenzen F., Tocilj A., Zarivach R., Harms J., Gluehman M., Janell D., Bashan A., Bartels H., Agmon I., Franceschi F., Yonath A. **Structure of functionally activated small ribosomal subunit at 3.3 Å resolution.** *Cell* 2000, **102**:615–623.

17. Carter P.A., Clemons W.M., Brodersen D.E., Morgan-Warren R.J., Wimberly B.T., Ramakrishnan V. **Functional insights from the structure of the 30S ribosomal subunit and its interactions with antibiotics.** *Nature* 2000, **407**:340–348.

18. Srajer V., Teng T., Ursby T., Pradervand C., Ren Z., Adachi S., Schildkamp W., Bourgeois D., Wuff M., Moffat K. **Photolysis of the carbon monoxide complex of myoglobin: nanosecond time-resolved crystallography.** *Science* 1996, **274**:1726–1729.

19. Abad-Zapatero C., Abbott Laboratories; Westbrook E. and Strasser S., Argonne National Laboratory. Personal communication.

chapter 16

1. Words used by Dr. Steve Cusack at the opening of the Workshop at the EMBL Grenoble Outstation. 1996.

2. Atkins P.W. *The Periodic Kingdom: A Journey into the Land of the Chemical Elements.* Basic Books—A Division of Harper Collins Publishers: New York, 1995.

3. Cowie D.B., Cohen G.N. **Biosynthesis by *Escherichia coli* of active altered proteins containing seleniun instead of sulfur.** *Biochim Biophys Acta* 1957, **26**:252–261.

4. Hendrickson W.A., Horton J.R., LeMaster D.M. **Selenomethionyl proteins produced for analysis by multiwavelength anomalous diffraction (MAD): A vehicle fo r direct determination of three-dimensional structure.** *EMBO J* 1990, **9**:1665–1672.

5. Abstracts of the annual ACA meeting in St. Louis, Missouri, July 19–15, 1997.

6. Hendrickson W.A. **Analysis of protein structure from diffraction measurement at multiple wavelenghts.in structure determination with synchrotron radiation.** *Trans. ACA* 1985, **21**:11–21.

7. Helliwell J.R. *Macromolecular Crystallography with Synchrotron Radiation.* Cambridge University Press: Cambridge, Cambridge, UK, 1992.

chapter 17

1. Tanner M. *Wagner*. Princeton University Press: Princeton, New Jersey, 1996, p. 172.

   Tanner, Michael; *Wagner*. Copyright ©1996 by Princeton University Press. Reprinted by permission of Princeton University Press.

2. Newman E. *The Wagner Operas*. Princeton University Press: Princeton, New Jersey. 1991, p. 417.

3. Bauer H.H. *Scientific Literacy and the Myth of the Scientific Method*. The University of Illinois: Urbana–Chicago, 1994.

4. Campbell P. **Tales of the Expected.** *Nature* 1999, **402**:C5–C94.

   Foreword to special supplement of *Nature* on *Impact of Foreseeable Science*.

chapter 18

1. Bernal J.D. *Science in History*. Vol. 1–3. Pelican Books: London, UK, 1969.

2. McKee, A. *Dresden 1945: The Devil's Tinderbox*. E. P. Dutton, Inc.: New York, 1984.

3. Vonnegut, K. *Slaughterhouse-Five or The Children's Crusade. 1969*. 25th anniversary edition. Delacorte Press/Seymour Lawrence: New York, 1994.

4. Rawls R. **Blobel gets Nobel in Medicine for work on protein signaling.** *Chem & Eng. News* 1999, **77**:13–14.

5. Kemp M. **Science in culture. Inverted logic. Antonio Gaudi's structural skeletons for Catalan Churches.** *Nature* 2000, **407**:838.

6. Feder T. **German synchrotron light source may find new home somewhere in Middle East.** *Physics Today* August 1999, p. 54.

7. Feder T. **Middle East synchrotron project moves ahead.** *Physics Today* February 2000, p. 52.

8. Feder T. **Jordan will likely host Middle East synchrotron light source.** *Physics Today* June 2000, p. 51.

9. SESAME Web site: http://www.sesame.org.jo

10. *Mycoses* 1999, **42**(3):Cover page.

Issue is corresponding to the 5th Congress of European Confederation of Medical Mycology, ©1999 Balckwell Wissenchafts-Verlag GmbH, Berlin.

11. Pasteur L. **Le budget de la science.** *La Revue des cours scientifiques*, 1er février 1968, pp. 137–139.

chapter 19

1. Mahoney T. *The Merchants of Life. An Account of the American Pharmaceutical Industry.* Harper & Brothers: New York, 1959.

2. Koehler C.S.W. **How thalidomide was kept out of the U.S. market. Brief review of the History of the FDA.** *Modern Drug Discovery* June 2000, pp. 69-72.

3. Kempf D., Marsh K.C., Denissen J.F. et al. **ABT-538 is a potent inhibitor of human immunodeficiency virus protease and has high oral bioavalability in humans.** *Proc Natl Acad Sci* 1995, **92**:2484–2488.

4. Erickson J., Neidhart D.J., VanDrie J. et al. **Design, activity and 2.8 Å crystal structure of a *c*2 symmetric inhibitor complexed to HIV-1 protease.** *Science* 1990, **249**:527–533.

5. Cameron D.W., Heath-Chiozzi M., Danner S. et al. **Randomized placebo-controlled trial on of ritonavir in advanced HIV-1 disease.** *Lancet* 1998, **351**:543–549.

6. Haseltine W.A. **Beyond chicken soup.** *Sci American*, November 2001. pp. 56–63.

7. Winterson, Jeanette. *Written on the Body.* Vintage International, Vintage Books. A Division of Random House, Inc: New York, 1992, pp. 61–62.

Reprinted by permission of International Creative Management, Inc. Copyright ©1994 by Jeanette Winterson. First Published by Vin.

chapter 20

1. Gordon J.E. *The Science of Structures and Materials*. Scientific American Library. A division of HPHLP: New York, 1988, p.182.

   ©J. E. Gordon. Reprinted with permission from the state of J. E. Gordon.

2. Diamond Jared. *Guns, Germs, and Steel. The Fates of Human Societies*. W.W. Norton & Co, Inc.: New York, 1998.

3. Watson J.D. *The Double Helix. A Personal Account of the Discovery of the Structure of DNA*. W. W. Norton and Co.: New York, 1980.

4. Judson H.F. *The Eighth day of Creation. Makers of the Revolution in Biology*. Jonathan Cape Ltd.: London, 1979.pp. 18-195.

5. Sayre A. *Rosalind Franklin and the DNA*. W.W. Norton & Company: New York, 1978.

6. Judson H.F. *The Eighth Day of Creation*. pp. 87.

7. Kroger et al. **Polycationic peptides from Diatom biosilica that direct silica nanosphere formation.** *Science* 1999, **286**:1059, 1129–1132.

8. Kamat S., Su X., Ballarini R., Heuer A.H. **Structural Basis for the fracture toughness of the shell of the conch *Strombus gigas*.** *Nature* 2000, **405**:1036–1040.

9. Ball P. *Made to Measure: New Materials for 21st Century*. Princeton University Press: Princeton, New Jersey, 1999.

10. Gross, M. *Travels to the Nanoworld: Miniature Machinery in Nature and Technology*. Perseus Publishing: Cambidge, Massachussetts, 1999.

11. Benyus J. *Biomimicry*. William Morrow & Co.: New York, 1977.

    "...a 380-million-year-old fiber with a twenty-first Century future."

12. Mason A. **The world of the Spider.** A Sierra Club—Books: San Francisco. 1999.

chapter 21

1. Hodgkin D. *A Life. Georgina Ferry*. Granta Books: London, 1998.

2. Perutz M. *Protein Structure: New Approaches to Disease and Therapy.* W.H. Freeman and Company: New York, 1992.

3. Chen L., Neidhart D., Kohlbrenner W.M., Mandecki W., Bell S., Sowadski J., Abad-Zapatero C. **3-D structure of a mutant (Asp101->Ser) of *E. coli* alkaline phosphatase with higher catalytic activity.** *Prot Eng* 1992, **5**(7): 605–610.

4. Dealwis C.G., Brennan C., Christianson K., Mandecki W., Abad-Zapatero C. **Crystallographic analysis of reversible metal binding observed in a mutant (Asp153->Gly) of *Escherichia coli* alkaline phosphatase.** *Biochemistry* 1995, **34**:13967–13973.

5. *Protein Engineering. Tutorials in Molecular and Cell Biology.* Oxender, Dale L., and Fox, C. Fred, eds. Alan R. Liss, Inc.: New York, 1987.

6. Bryson, J.W., Betz S.F., Lu H.S., Suich D.J., Zhou H.X., O'Neil K.T., DeGrado W.F. **Protein Design, a hierarchic approach.** *Science* 1995, **270**: 935–941.

7. DeGrado W. F. **Proteins from Scratch.** *Science* 1997, **278**:80–81.

9. DeGrado W.F., Summa C.M., Pavone V., Nastri F., Lombardi A. *De novo* **design and structural characterization of proteins and metalloproteins.** *Ann Rev Biochem* 1999, **68**:779–819.

10. Galli-Taliadoros L.A., Sedwick J.D., Wood S.A., Korner H. **Gene knock-out technology: A methodological overview for the interested novice.** *J Immunol Meth* 1995, **181**(1):1–15.

11. McHughen A. **Pandora's Picnic Basket. The Potential and Hazards of Genetically Modified Foods.** Oxford University Press: Oxford, UK, 2000.

12. *The DNA Story. J. D. Watson & John Tooze. A Documentary History of Gene Cloning.* W. H. Freeman and Company: New York, 1981.

13. Boutilier N. **No Justice.** In: *According to Her Contours.* Black Sparrow Press: Santa Rosa, California, 1992, p. 209.

14. Kim E.E., Wyckoff H.W. **Reaction mechanism of alkaline phosphotase based on crystal structures. Two-metal ion catalysis.** *J Mol Biol* 1991, **218**:449.

chapter 22

1. Crick F.H.C., Watson J.D. **Structure of small viruses.** *Nature* 1956, **177**:473–475.

2. Caspar D.L.D., Klug A. **Physical principles in the construction of regular viruses.** In: *Cold Spring Harbor Symposia on Quantitative Biology XXVII.* Cold Spring Harbor Laboratory: New York, 1962, **27**:1–24.

3. Harrison S.C., Olson A.J., Schutt C.E., Winkler F.K., Bricogne G. **Tomato Bushy Stunt Virus at 2.9 Å resolution.** *Nature* 1978, **276**:368–373.

4. Abad-Zapatero C., Abdel-Meguid S.S., Johnson J.E., Leslie A.G.W., Rayment I., Rossmann M.G., Suck D., Tsukihara T. **Structure of Southern Bean Mosaic Virus at 2.8 Å resolution.** *Nature* 1980, **286**:33–39.

5. Liljas A. et al. **Structure of Satellite Tobacco Necrosis Virus.** *J Mol Biol* 1982, **195**(1):93–108.

6. Grimes J.M. et al. **The atomic structure of Bluetongue Virus core.** *Nature* 1998, **395**:470–478.

7. Reinisch K.M., Nibert M.L., Harrison S.C. **Structure of Reovirus core determined via X-ray crystallography.** *Nature* 2000, **404**:960–967.

8. Perutz M. **From a Tomato Virus to Tumor and Influenza Viruses.** Chapter 8. In: *Protein Structure. New Approaches to Disease and Therapy.* W.H. Freeman and Company: New York, 1992.

9. Rossmann M.G., Abad-Zapatero C., Murthy M.R.N., Liljas L., Jones T.A., Strandberg B. **Structural comparisons of some small spherical plant viruses.** *J Mol Biol* 1983, **165**:711–736.

chapter 23

1. Kuhn T.S. *The Structure of Scientific Revolutions.* 2nd enlarged ed., Chicago University Press: Chicago, Illinois, 1970.

2. Polanyi, M. **The republic of science: Its political and economic theory.** *Minerva* 1962, 1:54–73.

3. Bauer H.H. *Scientific Literacy and the Myth of the Scientific Method.* University of Illinois Press: Urbana–Chicago, 1994.

4. Kuhn T.S. *The Structure of Scientific Revolutions.* 2nd enlarged ed., Chicago University Press: Chicago, Illinois, 1970, p. 10.

5. Judson H.F. *The Eighth day of Creation. Makers of the Revolution in Biology.* Jonathan Cape Ltd.: London, 1979.

6. Srajer V., Teng T., Ursby T. et al. **Photolysis of carbon monoxid complex of myoglobin: Nanosecond time-resolved crystallography.** *Science* 1996, **274**:1726–1729.

7. Yoshikawa S., Shinzawa-Itoh K., Nakashima R. et al. **Redox-coupled crystal structural changes in bovine heart cytochrome *c* oxidase.** *Science* 1998, **280**:1723–1729.

8. Balabin Y., Onuchi J.N. **Dynamically controlled protein tunneling paths in photosynthetic reaction centers.** *Science*, 2000, **290**:114–117.

chapter 24

1. Winfree A.T. *The Geometry of Biological Time. Biomathematics Monograph No. 8.* Springer-Verlag: New York–Heidelberg–Berlin, 1980.

2. Winfree A.T. *The Timing of Biological Clocks.* (Scientific American Library No. 19.) W. H. Freeman and Company: New York, 1987.

3. **A remarkable year for clocks.** *Science* 1998, **282**:2157

4. **The universe of *Drosophila* genes.** Jasny, B. ed. Special section of *Science* 2000, **287**:2181–2234; Adams M.D. **The genome sequence of *Drosophila melagonaster*.** *Science* 2000, **287**:2185–2195.

5. Winfree A.T. *The Geometry of Biological Time*. 2nd edition. Springer-Verlag: New York, 2001.

chapter 25

1. Prigogine I., Nicolis G., Babloyantz A. **Thermodynamics of evolution.** *Physics Today*, 1972, November and December issues.

2. Prigogine I. *Introduction to Thermodynamics of Irreversible Processes*. Wiley: New York, 1961.

3. Glansdorff P., Prigogine I. **Non-equilibrium thermodynamics.** In: *Encyclopedia of Physics*. 2nd edition. Rita G. Lerner and George L. Trigg, eds. VCH Publishers, Inc.: New York, 1991. pp. 1256–1263.
    Summary of the field of Nonequilibrium Thermodynamics.

4. Landauer R. **Stability in the dissipative steady state.** *Physics Today* November 1978.
    Referred to by Ball, Philip, in *The Self-Made Tapestry: Pattern Formation in Nature*. Oxford University Press, Oxford, UK, 1999.

5. Babloyantz A. *Molecules, Dynamics and Life*. John Wiley & Sons: New York, 1986.
    This is a fascinating little book. Very clear and accessible.

6. Epstein I.R. **Patterns in time and space generated by chemistry.** *Chem & Eng News* March 30, 1987, pp. 24–36.

7. Babloyantz A. *Molecules, Dynamics and Life*. Ibidem p. 172.
    Copyright © 1986 by John Wiley & Sons, Inc. Reprinted by permission of John Wiley & Sons, Inc.

8. Fee M.S. et al. **The role of nonlinear dynamics of the syrinx in the vocalizations of the songbird.** *Nature* 1998, **395**:66–71.

9. Capra F. *The Web of Life: A New Scientific Understanding of Living Systems*. Anchor Books: Doubleday, New York, 1996.

10. Eigen M., Schuster P. *The Hyper Cycle. A Principle of Natural Self-Organization*. Springer-Verlag: Berlin, 1979.

11. Kauffman, S.A. *At Home in the Universe: The Search of the Laws of Self-Organization and Complexity.* Oxford University Press: New York–Oxford, UK, 1995.

12. Lewontin R. *The Triple Helix. Gene, Organism, and Environment.* Harvard University Press: Cambridge, Massachusetts, 2000.

Reprinted by permission of the publisher from *The Triple Helix: Gene, Organism and Environment* by Richard Lewontin, pp. 113-114, Cambridge, Mass.:Harvard University Press, Copyright ©1986 Gius Laterza & Figli Spa, Copyright ©2000 by the President and Fellows of Harvard College.

13. MacKay, Brown G. *Selected Poems 1954-1992.* John Murray Publishers Ltd.: London, UK, 1996.

Reprinted with permission from 1996 edition. John Murray.(Publishers) Ltd. Albemarle Street, London. Introduction.

14. Winfree A.T. *The Timing of Biological Clocks.* (Scientific American Library No. 19.) W. H. Freeman and Company: New York, 1987.

chapter 26

1. Cole et al. **Diciphering the biology of Mycobacterium tuberculosis from the complete genome sequence.** *Nature* 1998, **393**:537–544.

2. The Genome International Sequencing Consortium. **Initial sequencing and analysis of the human genome.** *Nature* 2001, **409**:860–921.

3. Venter J.C. et al. **The sequence of the human genome.** *Science* 2001, **291**:1304–1351.

4. Fleischman R.D. et al. **Whole-genome random sequencing and assembly of *Haemophilus influenzae* Rd.** *Science* 1995, **269**:496–512.

5. Blattner F.R. et al. **The complete genome sequence of *Escherichia coli* K-12.** *Science* 1997, **277**:1453–1474.

6. Kunst F. et al. **The complete genome sequence of the Gram-positive bacterium *Bacillus subtilis*.** *Nature* 1977, **390**:249–256.

7. Goffeau A. et al. **The yeast genome directory.** *Nature* 1997, **387** Suppl:1–105.

8. **The universe of *Drosophila* genes.** Jasny B., ed. *Science* 2000, **287**:2181-2234; Adams M.D. **The genome sequence of *Drosophila melagonaster*.** *Science* 2000, **287**:2185–2195.

9. The *Arabidopsis* Genome Initiative. ***A. thaliana* genome.** *Nature* 2000, **408**:791–826.

10. Hoch J.A., Losick R. **Panspermia, spores and the *Bacillus subtilis* genome.** *Nature* 1997, **390**:237–238.

11. Chayen N.E., Helliwell J.R. **Protein crystallography: The human genome in 3-D.** *Physics World* 1998, **11**(5):43–48.

12. Pennisi E. **Structure of key cytoskeletal protein tubulin revealed.** *Science* 1998, **279**:978–979.

13. Szent-Györgyi A. *Introduction to Submolecular Biology.* Academic Press: New York, 1960.

chapter 27

1. Dawkins R. *Unweaving the Rainbow: Science, Delusion and the Appetite for Wonder.* Houghton Mifflin Company: Boston–New York, 1998.

Textbooks and Study Guides

bnsic chemistry

Most of the world around us is made up of chemical entities. A few starters.

1. Salem L. *Marvels of the Molecule.* VCH Publishers, Inc. New York, 1987.

   Translation from French of: *Molécule, la merveilleuse.* L. Salem.

   A delightful little book about the internal structure of atoms and molecules in very simple and descriptive language. Distributed worldwide by VCH Verlagsgesellschaft mbH P.O. Box 1260/1280, D-6940 Weinheim. Germany.

2. Atkins P.W. *Molecules.* Scientific American Library. A Division of HPHLP. New York. Disbributed by W.H. Freeman and Company. 1987.

   A lavishly illustrated book that gives a sense of what molecules are and why they are so important in our lives.

3. Atkins P.W. *The Periodic Kindom. A Journey into the Land of the Chemical Elements.* Basic Books. A Division of Harper Collins Publishers, Inc. 1995.

   A delightful exploration of the land and properties of the periodic table.

rocks and minerals

Although outside the main focus of this book, any book about crystals should have some references about minerals and gemstones.

1. Schumann W.. *Gemstones of the World*. Sterling Publishing Company, Inc.: New York, 2000.

2. SofianidesA.S., Harlow G.E. *Gems, Crystals and Minerals*. Simon & Schuster: New York, 1991.

biochemistry

You might like to consult some of the biochemical concepts discussed thorough the text in some extensive biochemical references.

1. Lehninger A.L., Nelson D.L., Cox M.M. *Principles of Biochemistry*. 2nd edition. Worth Publishing: Orange, Virginia, 1993.
   The successor to the classic *Biochemistry* by Albert L. Lehninger.

2. Stryer L. *Biochemistry*. 4th edition. W. H. Freeman and Company: New York, 1995.

protein structure

These references explain the fundamental principles of protein architecture. There is much to be discovered besides what is presented in what you read so far.

1. BrandenC, Tooze J. *Introduction to Protein Structure*. 2nd edition. Garland Publishing, Inc.: New York and London, 1998.

2. Schulz G.E., Schirmer R.H.. *Principles of Protein Structure*. Springer-Verlag: Berlin–New York–Heidelberg, 1979.

3. Lesk A.M.. *Protein Architecture. A Practical Approach*. IRL Press at Oxford University Press: Oxford, UK, 1991.

4. Dickerson R.E, Geis I. The Structure and Action of Proteins. Harper & Row, Publishers, New York, London. 1969.
   The classic book on protein structure.

crystallography and protein crystallography

These references explain the nuts and bolts of crystallography with different levels of mathematical complexity.

1. McRee D.E.. Practical *Protein Crystallography*. 2nd edition. Academic Press: San Diego, California, 1999.

2. Rhodes G. *Crystallography Made Crystal Clear*. 2nd edition. Academic Press: San Diego, California, 2000.
   A guide for users of macromolecular models.

3. Blundell T.L., Johnson L.N. *Protein Crystallography*. Academic Press: San Diego, California, 1976.

4. Drenth J. *Principles of Protein X-ray Crystallography*. 2nd edition. Springer-Verlag Inc.: NewYork, 1999.

5. *Crystallography Across the Sciences*. H. Schenk, ed. IUCr, 1998.
   A Celebration of 50 years of Acta Crystallographica and the IUCr. Reprinted from Acta Crystallographica A54, Part 6, Number 1. Edited by H. Schenk.

6. Pickworth J., Glusker J.P., Kenneth N. Trueblood. *Crystal Structure Analysis. A Primer*. 2nd edition. Oxford University Press: New York, 1985.

7. Stout G.H., Jensen L.H. *X-Ray Structure Determination. A Practical Guide*. 2nd edition. John Wiley & Sons: New York, 1989.

macromolecular crystallography and synchrotron radiation

This reference is the best starting point for the use of synchrotron sources for the study of protein structure.

1. Helliwell J.R. *Macromolecular Crystallography with Synchrotron Radiation*. Cambridge University Press: Cambridge, Massachssetts, 1992.

crystallization of biological macromolecules

Obviously you need crystals to do crystallography. The references that follow are important to learn how to grow crystals of biological macromolecules.

1. Davey R., Garside J. *From Molecules to Crystallizers—An Introduction to Crystallization.* Oxford University Press: New York, 2000.
   A brief insightful book on the basic principles of crystallization.

2. *Protein Crystallization: Techniques, Strategies, and Tips. A Laboratory Manual.* Terese M. Bergfors, ed. (IUL Biotechnology Series No. 1.) International University Line, La Jolla, California, 1999.

3. McPherson A. *Crystallization of Biological Macromolecules.* Cold Spring Harbor Laboratory Press: Cold Spring Harbor, Ney York, 1999.

4. McPherson A. *Preparation and Analysis of Protein Crystals.* John Wiley & Sons: New York, 1982.

5. *Crystallization of Nucleic Acids and Proteins.* A. Ducruix and R. Giegé, eds. IRL Press at Oxford University Press: Oxford. UK, 1992.

non-equilibrium thermodynamics, biochemical oscillations and biological complexity

A wide range of books to give you a panorama of the fascinating fields of complexity and chemical and biological oscillators.

1. Kondepudi D., Prigogine I. *Modern Thermodynamics: From Heat Engines to Dissipative Structures.* John Wiley & Sons Ltd.: New York, 1998.
   A novel approach to thermodynamics that includes the results of equilibrium theory in a more general and broader context.

2. Babloyantz A. *Molecules, Dynamics and Life: An Introduction to Self-Organization of Matter.* (Wiley Series in Nonequilibrium Problems in the Physical Sciences and Biology. Vol. IV. I. Prigogine and G. Nicolis, series eds.) John Wiley & Sons: New York, 1986.

3. Ricard J. *Biological Complexity and the Dynamics of Life Processes.* (New Comprehensive Biochemistry Series, Vol. 34, G. Bernardi, ed.) Elsevier: Amsterdam, Netherlands, 1999.

4. Nicolis G., Prigogine I. *Self-Organization in Non-Equilibrium Systems. From Dissipative Structures to Order through Fluctuactions.* John Wiley & Sons: New York, 1977.

5. Glansdorff P., Prigogine I. *Thermodynamic Theory of Structure, Stability and Fluctuations.* John Wiley & Sons, Ltd.: New York, 1971.

6. Goldbeter A. *Biochemical Oscillations and Cellular Rhythms.* Cambridge University Press: Cambridge, Massachussetts, 1996. The molecular bases of periodic and chaotic behavior.

7. Prigogine I. *From Being to Becoming. Time and Complexity in the Physical Sciences.* W. H. Freeman and Company: New York, 1980.

8. Kauffman S.A. *Origins of Order: Self-Organization and Selection in Evolution.* Oxford University Press: New York, 1993.

9. Kauffman S.A. *At Home in the Universe: The Search of the Laws of Self-Organization and Complexity.* Oxford University Press: New York, 1995.

10. Kauffman S.A. *Investigations.* Oxford University Press: New York, 2000.

World Wide Web Sites of Interest

Addresses and content of Web sites can change without notice. The sites provided contain many invaluable links.

1. International Union of Crystallography (IUCr): www.iucr.org.

   Valuable cross links to books and matters of interest to crystallography and crystallographers.

2. Advanced Photon Source (APS): www.aps.anl.gov

   The Advanced Photon Source published in 1997 in collaboration with the Nobel Foundation a small booklet with the biographies and contributions of scientists connected with X-ray science who received the Nobel Prize. Copies can be obtained from the user office.

3. European Synchrotron Radiation Facility (ESRF): www.esrf.fr

4. Nobel Foundation: www.nobel.se

   For information regarding scientists whose work was connected with X rays.

5. Arts Science Research Laboratory: www.artscienceresearchlab.org

6. Protein Data Bank (PDB):

   Now operated by the Research Collaboratory for Structural Bioinformatics www.rcsb.org

7. Expert Protein Analysis System server of Swiss Institute of Bio-informatics (SIB, now including the original Swiss-Prot, protein sequence data base): http://www.expasy.ch

   This is an excellent resource about protein sequences and protein structure.

8. History of Mathematics:
   physics.hallym.ac.kr/reference/physicist/mathematician.html

9. History of Crystallography: www.lmcp.jussieu.fr/~soyer
   Including 200 years of crystallography in France (in French).

10. Minerals and Gems: www-ext.lmcp.jussieu.fr/mineraux
    This is an excellent site on minerals maintained by the Laboratory of Mineralogy and Crystallography of Paris.

11. Mathematical Society of America: www.maa.org

12. History of Physics: physics.hallym.ac.kr/reference/physicist/physicist.html

13. Biophysics-Physics of Life: www.elsevier.com/locate/physoflife

14. American Institute of Physics (AIP): www.aip.org
    Diffraction theory is part of physics.

15. American Chemical Society (ACS): www.acs.org
    Many things of interest to biologists.

16. American Crystallographic Association (ACA): www.hwi.buffalo.edu/ACA
    There is wealth of resources about crystals and crystallography, including teaching tools.

17. CrystaLinks: www.bmsc.washington.edu/
    Excellent technical resource on the technical aspects of crystallography. Includes teaching resources.

18. Polycrystal Book Service: www.polycbs.com
    Full selection of books about crystallography, crystallographers and related topics such as structural biology, material science, physics of condensed matter and many more at discounted prices.

19. National Synchrotron Light Source (NSLS): www.nsls.bnl.gov

20. Protein Kinases: www.sdsc.edu/kinases
    Special site dedicated to proteins involved in signal transduction.

21. The Human Genome Project: www.ornl.gov/hgmis
    Entry point a variety of sites related to the human genome project.

22. The Institute for Genomic Research: www.tigr.org

TIGR, the not-for-profit center offers a wealth of information on this Web site. Many links to unpublished microbial genomes (approximately 30) and to more than a hundred sequencing projects underway.

23. Genome Landmarks:

http://www.sciencemag.org/feature/data/genomes/landmark.shl

Landmark papers in genomic-related research. Sponsored and maintained by *Science* magazine (A publication of the American Association for the Advancement of Science AAAS.)

24. Metabolic Pathways Model: http://wit.mcs.anl.gov

A resource on the interconnection of the metabolic pathways of the different organisms for which the complete genome has been deciphered or is currently underway. Many useful links.

25. Transgenic mouse: www.med.umich.edu/tamc

www.bio.purdue.edu/courses/Koniecznylab/TMCF.html

An example of a site where you can order transgenic mice on line.

26. Biosafety and Biotechnology Information sites:

American Council on Science and Health: http://www.acsh.org/

United Nations Environment Program on Biosafety:

http://irptc.unep.ch/biodiv

National Agricultural Library (USA): http://www.nalusda.gov/ibc/

27. Structural Genomics Sites: http://s2f.carb.nist.gov

A site where studies on the proteins from the flu pathogen (*H. influenza*) are in progress.

28. Virus structure: http://mmtsb.scripps.edu/viper/viper.html

29. Howard Hughes Medical Institute: http://www.hhmi.org

30. A.T. Winfree: http://www.cochise.biosci.arizona.edu/~art

An excellent resource for information relating to biological clocks.

copyright credits

Page vi: PEANUTS reprinted with permission of United Feature Syndicate, Inc..

chapter 1

Plate 1.1: Stamp design ©1984 United States Postal Service. Reproduced with permission. All rights reserved. Written authorization from USPS is required to use, reproduce, republish, upload, post, transmit, distribute or publicly display this image.

Plate 1.2: M. C. Escher's *Depth* ©2002 Cordon Art B. V.—Baarn—Holland. Reprinted with permission. All rights reserved.

chapter 2

Figure 2.1: First published in *Lustingen Blattern* circa 1900. Reprinted from *Nature*.[2] Courtesy of Macmillan Publishers Ltd.

chapter 3

Figure 3.1: Courtesy of International Union of Crystallography.

Plate 3.1: Stamp Design ©1984 United States Postal Service. Reproduced with permission. All rights reserved. Written authorization from USPS is required to use, reproduce, republish, upload, post, transmit, distribute or publicly display this image.

chapter 5

Plate 5.1b: Reprinted with permission from *Journal of Applied Crystallography*.[6] Courtesy of the International Union of Crystallography.

chapter 6

Plate 6.1: The music copyright ©1926 by Union Musical Española Editores, Madrid, Spain. International copyright secured. All rights reserved. Reprinted by permission of Associated Music Publishers, Inc. (BMI). C.A-Z appreciates the assistance of Jim Sukowski (Abbott Laboratories' Creative Network) in the composition of this image.

chapter 7

Plate 7.1: The data for the 1 Å map calculation courtesy of Dr. V. Nienaber (previously of Abbott Laboratories).

Plate 7.2a: Courtesy of John Helliwell and co-workers.

Plate 7.2b: Courtesy of A. Podjarny, A. Joachimak, and co-workers.

chapter 9

Plate 9.1: Courtesy of Quintas Fotógrafos (Vitoria, Spain)—the top image and my friend Ignacio Hernando—the bottom image. The assistance of Jim Sukowski and Jeff Frye of Abbott Laboratories' Creative Network is appreciated.

chapter 10

Figure 10.1: Inset drawn by the author and published in *Journal of Molecular Biology*;[5] reproduced with permission, courtesy of Academic Press. Main image courtesy of Michael and Audrey Rossmann.

chapter 12

Plate 12.1: The image was created by Dr. David Goodsell of the Scripps Research Institute and is reproduced here with permission ©David S. Goodsell, 2001; Courtesy of David S. Goodsell.[13]

chapter 13

Plate 13.1: Reprinted with permission from *Protein Science*.[5] Courtesy of Protein Society.

chapter 14

Page 103: Paraphrased from the brochures of the Ettore Majorana Center for Scientific Culture. Courtesy of Dr. Paola Spadon and Dr. Ludovico Riva di San Severino. Also, *Encyclopaedia Britannica*, 1974 edition.

Plate 14.1: Reprinted from *Nature*.[13] Courtesy of Dr. Hans Deisenhofer and Macmillan Publishers Ltd.

Plate 14.2: Reprinted with permission from *Science*.[14] ©2000 American Association for the Advancement of Science. Courtesy of R. Stenkamp with co-workers and AAAS.

Plate 14.3b: Image courtesy of T. Tsukihara and co-workers.[7]

Plate 14.3c: Compound image courtesy of D. A. Stock, A. G. W. Leslie, and J. E. Walker.[8,9] The structure of the c10 fragment was obtained by Nuclear Magnetic Resonance methods by Mark Girvin and co-workers.[12]

Plate 14.3d: Structure of the Potassium channel from the work.[16] Image created using PDB set 1bl8.

chapter 15

Plate 15.1: Photo credits: Abbott Laboratories group.[19] Dr. Cele Abad-Zapatero, Abbott´ Laboratories; Dr. Ed Westbrook; and Ms. Susan Strasser, Argonne National. Laboratory User Program Administrator.

Advanced Photon Source, Bldg. 401-B1154. 9700 So. Cass Avenue, Argonne, IL. 60439.

Plate 15.1, bottom left: Courtesy of Ed Westbrook.

Plate 15.2: Courtesy of R. Fenner (Advanced Photon Source Public Affairs office).

chapter 16

Plate 16.1: Photos courtesy of Ms. Chantal Argoud and Dominique Cornuéjols of the ESRF Public Relations office.

chapter 17

Page 131: Tanner, Michael; Wagner,[1] p. 172. Copyright ©1996 by Princeton University Press. Reprinted by permission of Princeton University Press.

Plate 17.1: The image of the APS in the upper right corner appeared in the invitation card to the dedication of the Advanced Photon Source on May 1, 1996. (Courtesy of Argonne National Laboratory, Susan Barr Strasser.) The background illustration is reproduced with permission from a photograph entitled *Rheingold* by Ms. Beatriz Schiller (Copyright ©2001 Beatriz Schiller) of the 1990 production of *The Ring* by the Metropolitan Opera of New York, James Levine conducting.

chapter 18

Plate18.1a: Aerial photograph by J. Todó ©TAVISA, reprinted with permision.

Plate18.1b: Courtesy of the ESRF public relations office.

Plate18.1c: Courtesy of the APS users office and public relations.

Plate18.1d: ©1999 Blackwell Wissenchafts-Verlag GmbH, Berlin. Reprinted with permission. This image appeared on the Cover of *Mycoses* 1999, Vol.42, no.3. (special issue of 5th Congress of European Confederation of Medical Mycology) is entitled *"Wiederaufbau der Frauenkirche in Dresden: Fotomontage des Kuntslers und Fotographen Jorg Schoner."*[10]

chapter 19

Page 147: Winterson, Jeanette. *Written on the Body*,[7] pp. 61-62. Reprinted by permission of International Creative Management, Inc. Copyright © 1994 by Jeanette Winterson. First Published by Vin.

Plate 19.1: Courtesy of Kent Stewart of Abbott Laboratories.

chapter 20

Page 151: J.E. Gordon. *The Science of Structures and Materials*,[1] p. 182. ©J. E. Gordon. Reprinted with permission from the state of J. E. Gordon.

chapter 21

Page 167: Excerpt from **No Justice**. Copyright © 1992 by Nancy Boutilier. Reprinted from *According to Her Contours*,[13] p. 209. with the permission of Black Sparrow Press. (Santa Rosa, CA).

Figure 21.1: Courtesy of Cathy Brennan, Abbott Laboratories.

Plate 21.1: Image created by the author, using PDB entry 1alk.[14].

chapter 22

Figure 22.1: Created by the author, using images provided by S. Wilder and M. G. Rossmann.

Figure 22.2: Reprinted from *Journal of Molecular Biology*[9] with permission from Academic Press Ltd. ©Academic Press.

Plate 22.1, left: Collage of the flier that Audrey Rossmann prepared in calligraphy as an invitation to the party in celebration of the structure of SBMV.

Plate 22.1, right: The atomic model of the virus was the courtesy of my friend from our Purdue days, Keichi Fukuyama.

Plate 22.2a: Reprinted from *Nature*.[6] Courtesy of David Stuart and co-workers (Oxford University, UK) and Macmillan Publishers Ltd.

Plate 22.2b: Reprinted from *Nature*.[7] Courtesy Steve Harrison and co-workers (Harvard University) and Macmillan Publishers Ltd.

chapter 24

Plate 24.1: Reprinted with permission from reference,[2] p. 90. Courtesy of Arthur T. Winfree.

Plate 24.2: Reprinted with permission from reference,[2] p. 94. Courtesy of Arthur T. Winfree.

chapter 25

Page 195: Babloyantz A. *Molecules, Dynamics and Life*.[7] p. 172. Copyright © 1986 by John Wiley & Sons, Inc. Reprinted by permission of John Wiley & Sons, Inc.

Pages 198–199: Reprinted by permission of the publisher from *The Triple Helix: Gene, Organism and Environment*[12] by Richard Lewontin, pp. 113–114, Cambridge, Mass.:Harvard University Press, Copyright ©1986 Gius Laterza & Figli Spa, Copyright ©2000 by the President and Fellows of Harvard College.

Page 199: MacKay, Brown G. Selected Poems 1954–1992.[13] Reprinted with permission from 1996 edition. John Murray. (Publishers) Ltd. Albemarle Street, London. Introduction.

Plate 25.1: Reprinted with permission from *The Timing of Biological Clocks* (1987).[14] Courtesy of A.T. Winfree and Scientific American Library.

index

A

of macromolecule(s), biological, 185
of molecule(s), 225
biological, 183
organic, 65
atoms
with anomalous response—*see* anomalous scatterers
axis(es),
cell, 14
crystal, 19, 48, 190
lattice, 14, 25–26

ℬ

B19—*see* virus, parvovirus B19, human
Bach, J. S.—*see* Bach, Johann Sebastian
Bach, Johann Sebastian, 22, 135
Bailey, M., 101
Bairoch, A., 93
Barcelona, Spain,
La Sagrada Familia, 137
Belousov, B. P., 194
Belousov-Zhabotinskii reaction, 194–195
Berg, P.—*see also* Gilbert, W. and Sanger, F., 92
Berman, H.—*see also* Hamilton, W. and Sussman, J., 92, 94
Bernal, John D.—*see also* Crowfoot Hodgkin, Dorothy M., 41, 134
biomaterials
fingernail, 153
hair, 102, 153, 156, 226
horn, 153
marine shells, 157
silk, 20, 102, 153
wool, 102, 153–154, 156
Blobel, G., 136
Box-Cox transformation, 99–100
Bragg,
W. H.—*see* Bragg, William Henry
W. L.—*see* Bragg, W. Lawrence
W. Lawrence, 18-19, 36, 41, 65, 219
William Henry, 18, 36, 65, 153, 219
Bragg's
angle, 219
law, 18, 26–27, 218–219

Bravais, Auguste, 12–13
BZ reaction—*see* Belousov-Zhabotinskii reaction

𝒞

CaMV—*see* virus, cauliflower mosaic
capillary tube, 70, 219
sealed capillaries, 42
Carangeot, Arnould, 11
Carl Von Maria Linnaeus—*see* Linnaeus, Carl Von Maria
Caspar, D.—*see also* Klug, A., 171
CAT(s)—*see* Collaborative Access Team(s)
CCMV—*see* virus, cowpea chlorotic mosaic
chemoreceptor(s), 107
chirality-see also handedness, 42
clocks
biological, 187–191
chemical, 194
Cochran, W.—*see also* Sayre, W. and Main, P., 65
co-factor(s), 41, 57, 219
Collaborative Access Team(s) (CATs), 118–120
Combs of the Wind, the, 69–70
common salt—*see* NaCl
complementary perceptions of reality, 205
computer graphics, 84
conservative structure(s), 192, 194–197
conservative/dissipative structures,
definitions, 194
inter-relation of, 197
copper sulfate
first diffraction pattern, 18
Córdoba, Spain,
Mezquita Cathedral, 49
Corey, R. B.—*see also* Pauling, L., 154, 156
CPMV—*see* virus, cowpea mosaic
Crick, F. H. C.—*see also* Watson, J. D., 70, 152, 161, 166, 170
Crowfoot Hodgkin, Dorothy M., 41, 76, 163–164
Crowfoot, Dorothy M.—*see* Crowfoot Hodgkin, Dorothy M.
crystal
axes, 14, 19, 24–25, 48

Ewald's
    inverse lattice, 29
    sphere, 26, 27
experiment as a dialog with Nature, 55–56,
    213

f

FDA—see Food and Drug Administration
FHV—see virus, Flockhouse
fingernail—see biomaterials, fingernail
flows/forces role in maintaining life, 203
Food and Drug Administration (FDA), 143
Fourier
    analysis, 30
    combination, 34, 64
    Jean-Baptiste Joseph, 32, 34–36, 58, 124
    summations, 58, 65
    sums, 34, 63–65, 86
    terms, 63–64, 66, 125
Fourier's theorem, 34–35
fractal
    geometry, 100
    number, 100
Franklin, R. E.—see Franklin, Rosalind E.
Franklin, Rosalind E., 152
Frauenkirche,
    Dresden, Germany, 136
frequency
    distribution, 98
    intervals, 222–223
    of a crystal—see frequency of a wave
    of a musical note—see frequency of a wave
    of a wave, 31, 34, 70, 220, 227, 229–230
    of oscillation—see frequency of a wave
    of the radiation—see frequency of a wave
Friedrich, W.—see also Knipping, P., 18
Frodo—character of Tolkien's trilogy, 82
FRODO commands, 84–86

G

Galilei, Galileo, 56, 78
gap-junction(s), 106

Gassman, J.,—see also Huber, R., 84
gem(s), 9, 103
gemstone(s), 4, 9, 41
    aquamarines, 39
    diamonds, 9, 39
    emeralds, 39
    opals, 39
    rubies, 39
    sapphires, 39
    topazes, 39
genetic engineering, 163, 165–166
    experiments, 168
genetic engineers, 159, 161, 165–167
Geneva, Switzerland,
    Swiss-Prot databank, 93
genome, 92, 95, 160–161, 191, 200–202
    human, xiv, 56, 95, 200
    of Arabidopsis thaliana, 95, 201
    of B. subtilis—see genome, Bacillus
        subtilis
    of Bacillus subtilis, 201
    of bacteria,
        circular, 160
    of Caenorabditis elegans, 95
    of Drosophila melanogaster, 95, 191,
        201
    of E. coli—see genome, Escherichia
        coli
    of Escherichia coli, 160–131, 202
    of Haemophilus influenza, 95
    of Homo sapiens—see genome, human
    of Mycobacterium tuberculosis, 200
    of Saccharomyces cerevisiae, 95
    of Vibrio cholerae, 95
    sequence,
        complete, 200–201
genome(s),
    complete, 95, 191, 200–201
genomics, 200
    structural, 95, 121, 200–201
Gilbert, W.—see also Berg, P. and Sanger, F.,
        92
goniometer, 11
    contact, 11
    precision, 11
Gould, S. J.—see Gould, Stephen Jay
Gould, Stephen Jay—see also Shearer,
        Rhonda Roland, 23, 28

P

Pasteur, L.—*see* Pasteur, Louis
Pasteur, Louis, 43, 56, 139
pattern(s) of protein,
    folding, 154
Patterson
    Laboratories, 192
    methods, 74
Patterson, A. Lindo—*see also* Harker, D., 74
Pauling, L.—*see* Pauling, Linus
Pauling, Linus,—*see also* Corey, R. B.,
    153–156
PDB—*see* Protein Data Bank
Perutz, M. F.—*see also* Kendrew, J., 66–67,
    76, 97–98, 183, 224
phase problem, 63, 224–225
    as 'Gordian knot' of crystallography, 63
photoreceptor(s), 107
Plato's caves,
    electronic, 95
Plato's images, 95
Pliny the Elder, 9–10
    and the definition of crystal, 9–10
Polanyi, M.—*see also* Kuhn, T., 181
Poliovirus—*see* virus, Polio
Prigogine, I.—*see* Prigogine, Ilya
Prigogine, Ilya, 192–195
Protein Data Bank (PDB), xiii, xviii, 91–95,
    120-121, 145, 147, 185
protein
    aggregate, 105
    architecture, 156
    chain(s), 67, 85-86, 153-154, 170–171,
        225
    chemistry, 86
    class, 105, 156
    complex(es), 172
    crystallography, xvi, 43, 54, 59, 66–67,
        71, 75, 84, 117, 127, 144, 163,
        165, 169, 181, 183, 188–189
    design, 165
    engineering, 119, 159
        programs, 119
    engineers, 159
        tools of, 159–163
    fold(s), 77, 201, 207
    folding, 121
    model(s), 82, 88, 101, 220
    molecule(s), 43–44, 57, 59–60, 74, 101,
        105, 144, 194, 217, 220–221, 226
    shape, 146
    size, 99
    structure(s), 58, 66–67, 70–71, 73–75,
        87, 92–93, 95, 97–98, 100–101,
        105, 107, 110, 116, 119–120, 125,
        127, 153, 163, 169, 183–184, 188,
        206, 223–224
    subunit(s), 105, 171
    surface(s), 98, 100–101
    synthesis, 121
    texture, 102
protein(s), xiv, xv, 42, 43, 57–59, 66, 70–71,
    73–76, 91–93, 95, 97–99, 101,
    103–105, 109, 121–122, 126, 146,
    153, 156–157, 160, 162–166,
    170–172, 183–184, 191, 194, 201,
    219, 220, 222, 225–228
    building blocks of, 154
    dimeric, 144
    fibrous, 97
    encoded, 201
    G-, 108–110
    globular, 97-98, 100, 156
        vs. fibrous, 97
    ion-channel, 106, 108-109
    membrane, 106, 109–111
    membrane-associated, 105–106, 108,
        110, 146
    membrane-bound, 105–106, 146
    morphological, 156
    receptor, 106-109
    symmetrical arrangement(s) in, 74,
        144, 164
protein-cutting enzyme, 135
proteomics, 200

Q

quasi-symmetry—*see* symmetry, quasi-symmetry

R

radius of gyration, 99–100
reaction center, 104–106
real space, 24, 226
receptor molecule(s), 108–109
receptor(s), 107–110, 148, 157, 163
  G-coupled, 109–110
reciprocal space, 29
  coordinates, 33, 58
Recuerdos de Alhambra (musical piece), 52
reovirus—*see* virus, respiratory enteric orphan
resolution, 33, 58, 59, 73, 202
  high, 33, 58, 59, 67
  low, 58, 67
  map, 58
  medium, 121
rhinovirus—*see* virus, rhinovirus
rhinovirus 14, human—*see* virus, rhinovirus
  14, human
ribosome(s), 121–122, 227
  structure of, 122
Richards box, 83
Richardson, J., 76, 174
Robertson, J. M., 65–66, 76
Roentgen, Wilhem Conrad, 15–16
Romé de Lisle, J-B.—*see* Romé de Lisle,
  Jean-Baptiste Louis
Romé de Lisle, Jean-Baptiste Louis, 10–11
Röntgen, W. C.—*see* Roentgen, Wilhem
  Conrad
Rossmann, M. G., xv, 73, 77, 79, 173, 177,
  223
Rotavirus—*see* virus, Rotavirus

S

San Sebastian,
  Donostia, Spain, 107
Sanger, F.—*see also* Berg, P. and Gilbert, W.,
  92
Sayre, W.—*see also* Cochran, W. and Main,
  P., 65
SBMV—*see* Virus, Southern Beam Mosaic

scientific
  method, 183
  research as puzzle solving, 182
selenium, 71, 121, 125-126, 163, 224
SESAME project, Jordan, 138
sexangulum, 10
shape function, 58, 63
  as combination of multiple ways, 63
Shearer, R. R.—*see* Shearer, Rhonda Roland
Shearer, Rhonda Roland—*see also* Gould,
  Stephen Jay, 23
silk—*see* biomaterials, silk
single-residue mutations—*see also* site mutagenesis, 164
site mutagenesis—*see also* single-residue
  mutations, 163
Snow, C. P., 205
Snow's two cultures, 205
sodium chloride—*see* NaCl
solvent channels, 41
Sommerfeld, Arnold, 24
Sovel, D., 10
  and *Longitude,* 10–11
spatial counterpoint, 22, 24, 29
SpV4—*see* virus, Spiroplasm phage 4
statistical descriptors, 98–99
STNV—*see* Virus, Satellite Tobacco Necrosis
Strandberg, B. and co-workers—*see also*
  STNV, 170, 173
structural genomics—*see* genomics, structural
Structure-Aided Drug Design, 144–147
structure factor, 223
structure(s),
  crystal, 19, 36, 76, 105, 145, 187
  high-resolution, 105–106
  protein—*see* protein structure(s)
Stuart, D.—*see also* Virus, Blue Tongue, 173
sucrose, 65
  structure of, 65
surface-area/volume ratio, 99
Sussman, J.—*see also* Berman, H. and
  Hamilton, W, xvi, 92
Swiss-Prot data bank, Geneva, Switzerland,
  93
symmetrical arrangement(s) in protein(s)—
  see protein(s), symmetrical
  arrangement(s) in